AF843286

COLLECTIVE INTELLIGENCE FOR SMART CITIES

Intelligent Data Centric Systems

COLLECTIVE INTELLIGENCE FOR SMART CITIES

Series Editor
FATOS XHAFA

CHUN HO WU
Department of Supply Chain and Information Management, The Hang Seng University of Hong Kong, Shatin, N.T., Hong Kong

GEORGE TO SUM HO
Department of Supply Chain and Information Management, The Hang Seng University of Hong Kong, Shatin, N.T., Hong Kong

FATOS XHAFA
Universitat Politècnica de Catalunya (UPC), Barcelona, Spain

ANDREW W.H. IP
Department of Mechanical Engineering, University of Saskatchewan, Saskatoon, SK, Canada; Department of Industrial and Systems Engineering, The Hong Kong Polytechnic University, Kowloon, Hong Kong

REINOUT VAN HILLE
RideOnTrack Ltd., Geel, Belgium

ELSEVIER

ACADEMIC PRESS
An imprint of Elsevier

Academic Press is an imprint of Elsevier
125 London Wall, London EC2Y 5AS, United Kingdom
525 B Street, Suite 1650, San Diego, CA 92101, United States
50 Hampshire Street, 5th Floor, Cambridge, MA 02139, United States
The Boulevard, Langford Lane, Kidlington, Oxford OX5 1GB, United Kingdom

ISBN 978-0-12-820139-8

For information on all Academic Press publications
visit our website at https://www.elsevier.com/books-and-journals

Publisher: Joseph P. Hayton
Acquisitions Editor: Kathryn Eryilmaz
Editorial Project Manager: Czarina Mae S. Osuyos
Production Project Manager: Selvaraj Raviraj
Cover Designer: Miles Hitchen

Typeset by STRAIVE, India

Dedication

Professor Xhafa dedicates this book to his brother, Lulezim, his wife, Aurora, and nieces, Alda and Bleona, for providing endless care to our late mother, Haxhire—Rest in peace!

Dr Wu would like to dedicate this book to his family and many friends. A special feeling of gratitude of his loving mother, Ivy whose words of encouragement. His sister, Wendy has never left his side. He also dedicate this book to his fiancée and many friends (from the Big Data Intelligence Centre) who have supported him throughout the process. He will always appreciate all they have done.

Contents

List of figures and tables

List of figures

List of tables

About the authors

Chun Ho Wu excels in his role with The Hong Kong Polytechnic University (PolyU) as a researcher with more than a decade of professional experience. Having held numerous roles with the university previously, his expertise precedes him. Over the years, Dr. Wu worked closely with many industrial partners to support and contribute to formulating their business strategies, operations improvement, and new product development. Dr. Wu received the Big Data Analysis Award from the China & Hong Kong Enterprise Market Development Association in 2014, the Best Reviewer Award for the journal entitled *Internet Research* in 2016, and the Top Peer Reviewer on Publons in 2019.

He regularly contributes to research papers on Industrial Internet of Things (IIoT), engineering optimization, and digital health. In collaboration with several scholars at PolyU, Tianjin University, Sun Yat-sen University, The University of York, etc., his project and research outcomes have been presented in 15+ international conferences and published in 70+ international refereed journals such as *Information Science*, *Internet Research*, *International Journal of Production Research*, *IEEE Communication Letters*, *Enterprise Information Systems*, *Expert Systems with Applications*, and *Industrial Management and Data Systems*.

Dr. Wu is a member of IEEE and a senior member of the Hong Kong Society of Quality. In addition, he serves on the editorial boards of various international journals under Taylor & Francis, SAGE, IGI Global, and MDPI.

George To Sum Ho is currently an associate professor in the Department of Supply Chain and Information Management at The Hang Seng University of Hong Kong (HSUHK). He received his PhD in the Department of Industrial and Systems Engineering (ISE) from the Hong Kong Polytechnic University (PolyU). His research areas focus on big data analytics, the Internet of Things (IoT), and smart city applications. He has published more than 70 international journal papers in SCI/SSCI-listed journals such as the *International Journal of Production Economics*, *International Journal of Production Research*, and *Expert Systems with Applications*. He has a score of 25 h-index for publications. He has also managed several ITF projects, collaborative projects, and consultancy projects with different commercial companies and organizations, namely MTR Corporation Limited, Pok Oi Hospital, and Sino Parking Services Limited. Besides, his specialty has allowed him to serve as the associate editor for the *International Journal of Engineering Business Management*.

Fatos Xhafa, PhD in Computer Science, is a full professor in the Technical University of Catalonia (UPC), Barcelona, Spain. He has held various tenured and visiting professorship positions. He was a visiting professor at the University of Surrey, United Kingdom (2019/2020), visiting professor at the Birkbeck College, University of London, United Kingdom (2009/2010), and a research associate at Drexel University, Philadelphia, United States (2004/2005). He was a distinguished guest professor at Hubei University of Technology, China, for 3 years (2016–19). Prof. Xhafa has widely published in peer-reviewed international journals, conferences/workshops, book chapters, edited books, and proceedings in the field (H-index 55). He was awarded teaching and research merits by the Spanish Ministry of Science and Education, by IEEE conferences, and has received best paper awards. Prof. Xhafa has an extensive editorial service experience. He is founder and editor-in-chief of *Internet of Things* (an Elsevier journal; Scopus and Clarivate WoS Science Citation Index) and of *International Journal of Grid and Utility Computing* (Emerging Sources Citation Index) and associate editor/editorial board member of several indexed international journals. He is founder and editor-in-chief of two books series: the Springer Book Series Lecture Notes in Data Engineering and Communication Technologies (Scopus, EI Compendex, ISI WoS) and the Elsevier Book Series Intelligent Data-Centric Systems (Scopus, EI Compendex). Prof. Xhafa is a member of IEEE Communications Society, IEEE Systems, Man & Cybernetics Society, and founder member of Emerging Technical Subcommittee of Internet of Things.

His research interests include IoT and cloud-to-thing continuum computing, massive data processing and collective intelligence, optimization, security and trustworthy computing, and machine learning, among others. He can be reached at fatos@cs.upc.edu. Also see http://www.cs.upc.edu/~fatos/ and at http://dblp.uni-trier.de/pers/hd/x/Xhafa:Fatos.

Andrew W.H. Ip has more than 30 years of experience in teaching, research, education, industry, and consulting. He received his PhD from Loughborough University (United Kingdom), MBA from Brunel University (United Kingdom), MSc in Industrial Engineering from Cranfield University (United Kingdom), and LLB (Hons) from the University of Wolverhampton (United Kingdom). He is now professor emeritus of mechanical engineering at the University of Saskatchewan and principal research fellow in the Department of Industrial and Systems Engineering of The Hong Kong Polytechnic University. He previously held various positions as associate professor and assistant professor in the same department, and associate director of Integrated Graduate Development Scheme (IGDS) up to 2017.

Prof. Ip and his research team were awarded "Gold Medal with the Congratulations of Jury" and "Thailand Award for Best International Invention from The National Research Council of Thailand (NRCT)" at the 43rd International Exhibition of Inventions of Geneva. And in 2013, he obtained the Natural Science Award (Second Class) presented by the Ministry of Education, China.

Prof. Ip has published over 270 papers with over 200 papers in SCI-indexed journals and has written multiple books and invited book chapters. He is also the editor-in-chief of *Enterprise Information Systems*, editor-in-chief and founder of the *International Journal of Engineering Business Management*, and an editorial board member of various international journals.

Reinout Van Hille is currently active as a VOIP software engineer in the private industry. He obtained his master's in electronics and ICT engineering technology at the University of Antwerp. His MS dissertation was completed during an Erasmus exchange at the Universitat Politècnica de Catalunya. This dissertation and associated journal paper focused on analyzing the various architectural layers within the IoT archetype and how the new concepts and theories of data enrichment may be applied within the IoT.

Acknowledgments

The work described in this book was partially supported by matching grants from the University Grants Committee of the Hong Kong Special Administrative Region, China (RMGS Project No. 700026 & 700021), and a funded project under the Partnership Research Programme from the Innovation and Technology Fund of the Hong Kong Special Administrative Region, China (Project Code: PRP/034/20FX). The close collaboration and previous support from Sino Parking Services Limited and Towngas Telecommunications Company Limited are greatly appreciated.

Fatos Xhafa's work is supported by the Spanish Ministry of Science, Innovation and Universities, Programme "*Estancias de profesores e investigadores senior en centros extranjeros, incluido el Programa Salvador de Madariaga 2019,*" *PRX19/00155*. Work partially done at the University of Surrey, United Kingdom, on leave from the Technical University of Catalonia, BarcelonaTech. Discussions with Prof. Paul Krause on various interesting topics of this book are highly appreciated.

We express our gratitude to Dr. Valerie Tang, Dr. Paul Tsang, Mr. Edmund Wu, Mr. Samuel Yeung, Miss Yen Ng, and the Big Data Intelligence Centre for their assistance provided, and also Bernadine A. Miralles, Balasundaram Kavitha, and McLean Danielle of Elsevier for help with the book production of this book.

Introduction

The urbanization rate of the world's population is constantly increasing. In the 1950s, only 30% of the world's population lived in cities; by 2014, the rate of urbanization had increased to 54%, and the United Nations predicted that by 2050, this figure would rise to 66% (Yin et al., 2015). As more and more people will live in cities worldwide, planning within a city such as transportation, water supplies and sewerage systems, residential and office buildings, and education and health become essential to retain or increase people's standard of living. With the aim of improving people's standard of living, the smart city is becoming more and more popular as a solution worldwide. In a smart city, citizens' quality of life would be enhanced by utilizing new technologies. For instance, new technologies like the internet of things (IoT), big data analytics, and cloud computing add entirely new dimensions to different advanced data collection, data processing, data analysis, and data storage services. With the use of these advanced services, the performance of different fields in a city would be enhanced, e.g., health care, energy, transportation, and security. Thus, smart cities would become safe, secure, environmentally friendly, and have higher efficiency and effectiveness, thus improving citizens' quality of life (Letaifa, 2015).

Cities have a huge impact on the environment and play an important role in social and economic aspects worldwide (Albino et al., 2015). In other words, better performance of cities would result in a better world. Hence, the development of smart cities would be beneficial to the world. However, potential problems and challenges should be considered. Therefore, this book aims to study and explore research outputs in adopting such advanced technologies in developing smart cities from theoretical, empirical, and practical perspectives. The book comprises 11 chapters that provide a comprehensive overview of smart cities and discussions on new interdisciplinary research directions that have become possible in smart cities areas.

Chapter 1 covers the main concepts and models, processing, anomaly detection, applications, and challenges for understanding the properties of data streams. Applications of smart cities such as harbor planning and waste management with data streams in the Internet of Things (IoT) are illustrated.

Chapter 2 introduces the basics of semantic web technology and then the concepts of semantic data streams and semantic stream processing are discussed as topics of interest in the context of smart cities and collective intelligence.

Chapter 3 presents an overview of smart cities, including definitions, characteristics, measurement, and practices. With IoT and big data analytics as a fundamental component, smart cities implementations are discussed in detail to gain insights from the current practices to prepare for better future projects.

Chapter 4 reviews details and applications of IoT, cloud computing, and big data collection intelligence in smart cities. The benefits and challenges of smart cities development with the integration of IoT, cloud computing, and big data collection intelligence are discussed by providing an example of smart cities applications.

Chapter 5 presents a case of smart energy network application in smart cities using an example from China. An actual example of a smart energy network is introduced to show how smart energy networks use IoT solutions to provide a comfortable, convenient, safe, and energy-conserved environment to households and buildings.

Chapter 6 introduces a case of a smart firefighting system to achieve safety and effectiveness in firefighting and fire protection. A Chinese smart firefighting IoT platform is shown as an example to demonstrate the concept and application of smart firefighting and fire protection.

Chapter 7 illustrates the case of a smart car park. With the integration of RFID, WSAN, NFC, IoT, and cloud computing, an IoT-based advanced automobile parking navigation platform (IAPNP) is designed to solve the traffic handling and transportation problems in Hong Kong.

Chapter 8 presents the case of a smart crane selection system (SCSS) with the pilot study in the case company as an example. The operation performance in handling project cargo orders would be improved by SASS and hence increases the service satisfaction.

Chapter 9 discusses the TransAnalytics research project that aims to develop transport analytics, sustainable models, transport and citizens' mobility (T&M) algorithms, and software solutions that support stakeholders in the industry and local and national governments. Research problems and challenges in T&M are also discussed.

Chapter 10 discusses the use of blockchain to create the foundation of technology and information management to enhance the functionalities and reliability of connected systems in smart cities. Case sharing throws light on the background information and industrial criteria in various smart communities.

Finally, Chapter 11 presents the conclusions and future research directions for the smart city. With the rapid growth of advanced technologies applications in smart cities, it is believed that citizens' quality of life can be improved to a greater extent.

References

Albino, V., Berardi, U., & Dangelico, R. M. (2015). Smart cities: Definitions, dimensions, performance, and initiatives. *Journal of Urban Technology*, *22*(1), 3–21.

Letaifa, S. B. (2015). How to strategise SMART cities: Revealing the SMART model. *Journal of Business Research*, *68*(7), 1414–1419.

Yin, C., Xiong, Z., Chen, H., Wang, J., Cooper, D., & David, B. (2015). A literature survey on smart cities. *Science China Information Sciences*, *58*(10), 1–18.

CHAPTER ONE

Data streams—Concepts, definitions, models and applications in smart cities

1.1 Introduction

In the general sense, a data stream can be defined as an unbound sequence of time-varying, coherent data, which is in the process of being transmitted. The broad definition shows many of the typical properties of data streams, which should be understood very clearly as they form the fundamentals of many challenges related to processing data streams. Given the everywhere presence of data streams in all fields of applications, a large body of research results has been published in the literature (Ahmad et al., 2017; Calders et al., 2014; Kim & Park, 2020; Ramírez-Gallego et al., 2017; Xhafa et al., 2020). This chapter aims to lay the groundwork for understanding data streams, main concepts and models, processing, anomaly detection, applications, and challenges.

First and foremost—a data stream may be assumed to be unbounded. This does not imply by definition all data sources are assumed to produce data perpetually. Still, either starting point or ending point, or both will not be known in the majority of cases. This translates into storage capacity-related issues and the need to make design choices on processing the stream as it is ongoing.

Second, data in a stream is sequential by nature. Often the sequence of the data is essential and needs to be preserved. There are very few data streams for which the sequence is not crucial information. For this reason, timestamps or sequence numbers are added. These have additional benefits for handling error scenarios such as data loss or out of order data. From an information point of view, the sequence indicator should ideally be added to the data by the source. This indicator allows for the most accurate fault handling. The full timestamp also has an advantage over a raw sequence number, as it holds more information. Not all data sources produce data at a set rate, in

Collective Intelligence for Smart Cities
https://doi.org/10.1016/B978-0-12-820139-8.00001-2

which case the timestamp is very much preferred. However, from a cost point of view, the addition of the sequence indicator will induce an unavoidable overhead and thus require more bandwidth, and the timestamp is more expensive than the sequence number.

Data streams have a certain form of cohesion between elements in the stream. The cohesion between data elements, however, is an entirely abstract property. Said cohesion also typically becomes more complex as data travels further from the source, and this may be viewed as an increasing degree of heterogeneity. In the most simple scenario, the cohesion of the data is that of a single source transmitting a single type of information. This is a perfect homogenous stream. In modern-day applications, however, data sources can transmit multiple types of information, increasing the degree of heterogeneity. As data travels further upstream, multiple streams are often aggregated into a single stream. This creates complex streams and often identifies the data source of the elements within a stream.

As data turns out to be a precious resource, more and more devices are connected to the Internet each year. This idea of connecting all things to the Internet, including the resulting applications and challenges associated with it, has led the past decade to be marked by the paradigm known as the Internet of Things. This paradigm shift has also been a driving cause for innovations on all levels regarding data streams. On the sensing level, various smart sensors have been developed. On the processing level, developments were made regarding capacity, creating more performant clustering and cloud architectures. Various algorithms were developed for more intelligent processing at these high-capacity points.

Furthermore, the processing is expanded further and further toward the edge and fog layers (Alencar et al., 2020; de Assunção et al., 2018; Gomes et al., 2021; Xhafa et al., 2020), resulting in both edge processing and fog processing ideas, which have the primary goal of improving the reaction time of applications. On the networking level, the ever-increasing amount of data sources and their capabilities has exploded the amount of data transmitted. Matching this rapidly increasing need for faster data rates and better response times has led to developing the 5G networks.

The Internet of Things is now a part of our society in terms of numerous resulting applications. New applications are discovered regularly as the understanding of the capabilities of these technologies are better understood. Among these applications, for example, are the *real-time* information now common in public transport. Busses regularly update their GPS position to a server, which allows accurate estimates of the arrival time. Entire harbors have been fully automated by tagging containers, learning algorithms for automated planning, and even connecting the cranes to automated movement systems.

Not only harbor planning may be optimized, but also waste management. Waste bins equipped with sensors transmitting information regarding their contents allow for optimized and timely routing of the pickup trucks. Using knowledge of how full bins are in advance, a truck may decide to continue down streets it usually would not before returning to empty their cargo. In the case of general waste, the optimization may seem minimal as it generally accumulates at a more or less stable pace, but this is not always the case. For instance, the COVID-19 pandemic is causing pollution in different forms, such as air pollution, plastic pollution, etc. (Benson et al., 2021). It introduces a spike in waste that is less predictable due to tiny items used every day by the mass population. As another example, the accumulation of specific types of waste may be less predictable. This accumulation can also be the case for collecting used batteries or scrap metal, generally small recycled items.

1.2 Types of stream processing and levels of event processing

The goal is often to analyze the stream to find specific events of interest when processing streams. Such analysis may be differentiated chiefly based on the reactivity of the overall system. When no immediate actions are necessary, the data analysis may be done on a database (RDBS and/or NoSql) basis. A large portion of the data stream, or the whole stream, is recorded in a database before performing some analysis. Statistical analysis is often helpful to understand the distribution of the data and draw conclusions regarding the long-term behavior of the data. An example of this would, e.g., be an end-of-year analysis of the processing usage of a cloud infrastructure, which may be used to adjust provisioning needs.

If actions are necessary, the processing and analysis of the data are done *online (real-time)*. In online analysis, the system requires some action to be taken in reaction to specific data within a certain time frame. This allows data within a set window around the current data element in the sequence to be used for analysis. In some cases, the analysis of a data element (data tuple) may need to be done instantly. This is the case for applications that require certain critical reaction decision-making times. The main difference with the online approach is that the processing performed cannot include *future* samples of the current data for its computation. This complicates the analysis considerably, as data of interest is often dependent on its context, and samples partly define the context following the current element.

Typically the area of interest for online and instantaneous is detecting a particular event of interest and taking appropriate action. Events may be distinguished in terms of complexity by setting several criteria. The most simple of events, named atomic events, consist of singular simple value comparison. This type of event can be detected regardless of past and future data. They also require the least computational effort.

Increasing the complexity, composite events consist of a combination of different atomic events that must co-occur. Assuming a simple case, a single stream may be producing more complex data structures, e.g., a sensor stream of data structures containing temperature, pressure, and oxygen levels. Each data element may be evaluated to see if the set of events regarding temperature, pressure, and oxygen combined creates a dangerous situation. The event remains well definable as it is time-invariant, and only a single sample of the data stream needs to be assessed.

A far more prominent increase in complexity is added by involving the aspect of time. The time for one assumes that many more samples may be needed in the computation for detecting the event. This drastically increases computation time and memory requirements. Furthermore, the pattern that is sought after in time is not always a known one. Known patterns can still be found using relatively lower computational overhead based on a matching process. Events with a time complexity factor with defined event patterns typically involve data sources with clearly determined behavior. For data sources with less defined behavior, this becomes more complicated. For example—tracking the speed of a train to determine anomalous behavior is far more accessible as the expected behavior may be defined quite well depending on the segment. Attempting the same for a person is much more difficult as their expected speed is much more volatile.

Lastly, it should be noted that all sorts of event processing may be complicated more by the nature of the stream in other ways. Streams may inhibit an incoherent generation rate, making it much harder to determine the boundaries of events and how closely related events may be without the availability of basic timing information. In the simple example given for a composite event, the assumption was made that all data was generated in one structure. This is, in fact, a simplification, as in practice, the data may be divided across different streams with different generation rates. There is also the possibility of data loss and uncertainty regarding sensor accuracy.

1.3 Stream processing models

A distinction in the processing approach may be made based on how much of the data are used and what data are used for analyzing the current sample. Analysis of the current sample may be performed online, using only historical information. This maximizes responsiveness. Using only a subsection of samples by taking a window rather than using all historical samples lowers the computational overhead. However, certain events may not be detected without performing a historical analysis on a sample sometime later, which allows for *future* samples. Determining the occurrence of a turning point, e.g., a local minimum in weather temperature, cannot be achieved without historical analysis. The size of the window also limits what events may be detected. Events who are defined over a longer period cannot be detected using narrow windows.

Does the system need to adapt? Processing may update a set of parameters of the internal analysis algorithm based on incoming data. This may be done regardless of whether or not a window is used for processing. For many applications, the behavior of the data source being observed changes over time. If the data source changes, so should the analysis of its observed data.

Lastly, a topic of interest is the idea of sampling. In statistics, a subsample of a large data set is taken. This subsample is then used to draw certain conclusions regarding the data set as a whole. If the subsample taken is representative of the whole set, these conclusions are pretty accurate. Similar concepts may be applied to a data stream. Rather than processing the whole stream, only a few samples are extracted and used for processing. This may be done by occasionally processing a window or taking samples following a specific distribution or pattern. Regardless, the same conditions regarding the accuracy of the processing hold—the sampling must be performed so that the subset used for processing matches the behavior of the whole stream. This is harder for streams with less consistent behavior. As an example: consider a server mapping the coverage of the 5G network. It receives a datastream of perceived signal strengths from some clients, which generates a heat map. Sampling may be performed based on the distance traveled by the client using its GPS position. The resulting heat map may be assumed to be expected to be accurate if not for objects blocking the signals.

1.4 Anomaly detection methods from datastreams

The problem of finding anomalies or deviant data in a data-set is a topic in a broad range of applications, even outside of the streaming context. It has been studied for a long time, and a wide variety of different approaches exist. Some of these may be adapted to be used in stream processing. The various approaches can be split into several categories (Lindemann et al., 2021; Zhou et al., 2021).

1.4.1 Matching techniques

In cases where anomalies exhibit well-known patterns, the data can be matched to the patterns to have an effective yet computationally simple algorithm. The patterns, however, need to be known—which in a way asserts a problem with lower complexity. A threshold for matching accuracy needs to be applied with great care.

1.4.2 Statistical approaches

The oldest approaches to anomaly detection are the statistics-based approaches. Assuming the data-set is generated by a statistical model, the chance that a particular data element is obtained from it is known. Values that have a significantly low chance of occurring are considered outliers, which can be considered anomalies. In a Gaussian model, a simple method is to use a box-plot, which equates to considering any values that are not within 3σ from the mean to be anomalous. In a data streaming context, these kinds of methods offer a low computational complexity, but the data will not be generated following some statistical distribution in many cases.

1.4.3 Regressive approaches

Similar to the statistical approach, there are the regressive approaches. In these approaches, a regression model is fitted to the data. The difference between the actual data and the regression model, often called residual, can then determine how abnormal the behavior is. Once again, the main issue is that data are often not generated following a transparent model. A critical issue with both methods is that they require the source sets behavior not to change over time—as an outdated regression or distribution applied to a changing data-set would result in a considerable amount of false positives and negatives.

Nearest neighbor-based techniques build upon the idea that regular data instances have close neighbors, while anomalies occur far away. These algorithms use the distance to its k nearest neighbor or compute the density of the data around an instance. The local outlier factor technique is one of the significant contributions in this field. It compares the local density of an instance to that of its neighbors to detect anomalies. Nearest neighbor techniques are computationally more expensive than the previous methods but can be applied more often in the general case. They are also more resistant to changes in data nature over time.

Clustering techniques—Clustering techniques can be considered similar to density-based techniques because they often require a kind of distance computation to do the clustering. The critical difference is that each instance is compared to the cluster it belongs to rather than the local neighborhood. In the case of clustering-based techniques, an anomaly is either not part of a cluster, part of a sparse cluster, or should lie close to their respective cluster centroid. One of the main concerns for clustering-based anomaly detection is the high computational complexity.

Neural networks—These are more complex algorithms used for anomaly classification and prediction. The baseline for many of them is that they maintain a particular internal state. This state is typically trained initially: the system is fed a data source that exhibits the ideally expected behavior of the data, which may have been obtained through accurate simulation or by prerecording. The output of the system is then matched with the expected evaluation result to improve the internal state. When training is finished, the state will accurately match the desired system behavior and may be used to classify/predict anomalies.

1.5 Challenges of real-time stream processing: Big data vs big data streams

1.5.1 Characteristics of big data—The standard 5Vs

The IoT environment introduces several challenges. These are often tied to the challenges associated with big data. The challenges when dealing with big data are related to the "5Vs" of the characteristics used to identify big data. The big data Vs being:

Volume: It is the most commonly identified characteristic. The volume is the number of bytes of data needed to be processed and stored. This number rapidly expanded from gigabytes to terabytes and, for some applications, even petabytes. A popular web service has accumulated vast volumes of logs over the years. The decision is made to analyze these logs in hopes of reaching new conclusions. Processing these high volumes of data is the primary focus of big data analytics.

Velocity: As the amount of connected devices to the Internet grows, the data generation rate grows with it. The aspect of velocity refers to this generation rate. It demands the ability to transport and process high volumes of data within a short time frame. Every minute, hours worth of video content are uploaded to youtube by its users.

Variety: The variety refers to the unstructured nature of the data. Data generated by streams may not have a set in stone structure, making it harder to store in relational databases and further complicating processing. As an example, consider a stream of social media posts. Each post may or may not tag several other people, may or may not contain image or video content, the length varies, etc. The structure of a single post does not have a clear-cut definition. Relational databases were not designed to support the insertion of files for the video content.

Veracity: As loads of data are generated and processed from various sources, not all data will likely be accurate. This can occur due to transmission errors, faulty sensors, tampering, etc. This is the aspect of integrity. The accuracy of the data sources has a significant impact on the analysis results.

Value: Last but not least important. Although the raw scalability of the cloud has improved things, the previously mentioned characteristics combined require robust, expensive infrastructure. However, many applications have proven that such an investment is worth doing, provided the analysis of the data can create sufficient value.

Some developments were made to handle the challenges that come with these characteristics. The cloud and clustering infrastructures are most suitable for storing and processing high volumes of data. Network speeds both on wire and over the air have surged immensely. Various NoSQL types of datastores handle all sorts of use cases of unstructured and semistructured data. MapReduce, along with the general idea of splitting data into sets optimized for parallel processing, allows for more optimal usage of the newly available infrastructure.

1.5.2 Characteristics of IoT big data streams—The 5Vs revisited

The previous description of the V properties was mainly considering big data in a more static context. Here, the goal here is to have a form of analysis performed retroactively after storing the big data. This misses a few critical aspects related to the online processing of these big data streams. Revisiting the same Vs, but putting them in a more online, IoT streaming processing context adds further complexity to some of the Vs:

Volume: The volume of a data stream may be considered by its throughput, which relates more to the velocity in the static context. The volume of a data stream can be described as the total amount of data generated in a single period.

Velocity: In a static context, velocity refers to the rate at which the stored high volume expands due to data being added to it. While storing data is exciting and essential for many applications, in an online streaming context, the actual generation frequency (in Hz) of the various data sources and how responsive the system should be to each of the samples is vital.

Variety: In an IoT environment, many different data sources may be involved in a single application. This may be because the need to obtain different sorts of data exists. It is also possible many different flavors of a sensor exist that use the same type of data to partake in the system. A perfect example of the latter would be smartphones using a GPS application for tracking running performance. Each sensor has unique characteristics in terms of accuracy, frequency, fault tolerance, etc.

Veracity: This remains similar, but as the IoT range of applications extends, intelligent devices are used more and more in extreme circumstances. Applications are typically over the air, implying more loss of data.

Value: The need to create some sort of value through processing these data streams remains unchanged.

When processing big data streams, the main challenge is thus simple to formulate yet hard to solve: It is to evaluate and process the data in a timely and accurate fashion, so all events and data are handled and stored correctly. The first key issue here is that timely and accurate are somewhat contradictory. Increasing complexity requires more expensive processing algorithms to achieve better accuracy, requiring more processing time. Furthermore, the need for timely evaluation implies a part of the context regarding a data sample is lost. There is no (or limited) time to wait for *future* samples to elaborate on the context of the current sample.

The timely analysis must furthermore be placed in the context of a high input rate generated by vast amounts of data sources. Processing such high amounts of data requires the processing to be done in parallel across multiple cores or devices. This must be achieved without losing information. Many of the streaming processing platforms thus offer means for defining how the data stream may be divided across multiple processors and some form of duplicating the data to avoid data loss.

Accurate, fast, and online processing is hard to achieve. The machine learning and neural network processing techniques typically rely on continuous reevaluation of data as their model adapts from evaluating additional data. In a data stream application, the initial application response is often needed within a short time frame. This implies a sample can be viewed as to be evaluated *only once*. This type of processing is still relatively less established and thus does not yield the same accurate results compared to processing data from storage. Custom solutions often allow for more accurate results than more generic processing results but are typically less resistant to changes like the data source. Over time, the behavior of the data stream may change due to various reasons but mainly caused by changes in the behavior of the subject being observed. This problem is known as concept drift and is hard to handle, as it requires the model used for processing the data to adapt to it.

Retraining a model requires the availability of a more extensive set of data, i.e., there is a need to store the data on disk. For many applications, this will generically provide some form of historical information and allow the previously mentioned reevaluation of the data. Although the initial processing evaluation is essential, it does not exclude the possible use of later reevaluation for correcting the evaluations. This would at very least allow notification at a later stage when an incorrect evaluation has been made.

1.5.3 A new challenge: Anonymity and privacy

Over time as the IoT expands, humanity has begun to become a part of it more and more. A new challenge paired along with this increased presence is the preservation of privacy and anonymity of sensitive data in data streams (Chamikara et al., 2019; Loukides & Gkoulalas-Divanis, 2012; Nikas et al., 2018; Sopaoglu & Abul, 2020; Stephanie et al., 2021). When data are transmitted to the Internet, it may contain personal information. Often the critical rule in privacy is as follows: the data may be gathered and stored, but it should be impossible to link this data to a person or an entity. This is called the data anonymization process. As a recent example, anonymization was vital in the design of many of the "Corona virus Apps." These apps had to be designed so test results could not be linked to personal IDs.

However, in this case, transmitting personal information can be avoided relatively easily. A more complicated example is that of a live stream in a non-public area. Principally, the official permission of each person recorded in the footage should be acquired before sharing on the Internet. The GDPR has a separate document with examples regarding the suggested behavior (Rhahla et al., 2021). This may not always be possible, so alternatively, the faces of the individuals recorded in the background should be blurred automatically, without affecting the rest of the footage. The processing would need to keep up with the framerate of the camera footage and ideally have a minimal delay. Thus, anonymization can genuinely add a high degree of complexity to an otherwise more simple process.

Encryption, however, also provides a means of privacy and anonymity. The popular app Whatsapp shows a message indicating the end-to-end encryption of conversations, which advertises the increased privacy and protection. Many applications recognize the value of privacy and security and will only transmit encrypted data to servers. Servers still can decrypt the data, and typically the design of mobile applications requires users to consent for all sorts of permissions for sharing their data. This supports the ideas previously mentioned. However, a topic of interest might be to apply homomorphic encryption (Alloghani et al., 2019) to the data. In homomorphic encryption, the encryption allows the preservation of certain mathematical operations on the encrypted data (as of now limited). The obtained result when decrypting, even after performing these operations, thus remains valid. This way, the data could remain encrypted fully even when passed to the server, allowing only the client device to decrypt the result. This guarantees a high degree of anonymity for the user, but in general secure homomorphic encryption over streams is hard to perform.

1.6 Various levels of stream processing and their goals

1.6.1 Processing goals

In general, processing for various applications can be identified to follow a specific flow of processing tasks that need to be performed. The following processing tasks may be identified.

1.6.1.1 Filtering

In a basic filtering process of the data, there are two possible tasks to perform in this process. One task is filtering for data of interest, selecting the appropriate data attributes for the given context in multivariate data streams.

In a sense, filtering is the opposite process of event detection. It is the detection of when an event isn't occurring. The other task can be dealing with noise and erroneous data in uncertain environments.

1.6.1.2 Event detection

The event detection process may vary in complexity depending on the nature of the stream and the *definability* of the event. In the ideal case, events are of a particular type and easily detected. This could be the case, for example, when the event of interest is a certain sensor value exceeding a certain threshold value. This is a trivial detection in the case of a nonheterogenous stream. A simple query on the stream element to see if it exceeds the threshold value is sufficient. When the stream is heterogeneous, however, detecting the same event may be more tricky. The level of complexity depends on the type of data and the ordering. If all types of data in the stream are structurally different and thus may be identified uniquely, the problem remains trivial. If some are structurally equal, the system may rely on the ordering of the events if they are generated in a consistent pattern, e.g., event A always followed by event B.

The possibility of data loss further complicates this. If an instance of event A is lost and undetected, the system may mix up categorizing the events. Even without data loss, streams may not have predictable ordering due to having nonconsistent generation frequencies. A license plate scanner will trigger an event whenever a car passes, and cars do not pass at a consistent input rate.

In the most challenging case, it may not be possible to have a good definition of the events of interest. This is often the case for streams containing complex data. As an example, a hacked Twitter account may behave subtly differently from the user's normal behavior. Both the original data stream and the strange behavior are complicated to define.

1.6.1.3 Data enrichment

This is a topic of interest that is, in fact, rarely present in most live environments. The core idea of this process is to manipulate the representation of the data so that processing at later stages will benefit from it. This process involves labeling the raw data and attempting to add as much information as possible to it. This improves the processing of the data later on. An essential task in this regard is adding the generation time to the data for ordering purposes.

1.6.1.4 Data analysis

This may involve the computational process of finding more complex events if necessary. The goal of the processing is to *reason* with the data. This is the process of analyzing data for events and then using those events to draw new conclusions. Such conclusions may be tied to a set of actions.

1.6.1.5 Application processing

These are related to the set of actions previously mentioned. In general—the application may be considered the part of the processing that influences the natural world somehow.

1.6.2 Processing levels

Aside from the various processing tasks, a number of different processing levels may also be identified based on their physical location and architecture. These processing levels and their tasks are closely related. The processing levels identified are data sensing, preprocessing, edge/fog, cloud/clustering and finally, the client. When discussing the processing levels and their associated tasks, it's vital to consider the core parameters of available computational power and network bandwidth.

1.6.2.1 Data sensing

In any IoT architecture, the lowest level is data sensing (also referred to as Data generation). The generation layer is critical in determining the system's throughput characteristics. There is a data generation rate at this layer, which has to be accommodated in upper layers to ensure events are not lost. The frequency at which data are generated is a key parameter separate from the actual data rate. This frequency is a critical number, as any form of processing and storage of the data should be performed minimally at the same frequency to prevent discarding data.

1.6.2.2 Data preprocessing

The preprocessing layer is situated in the first computationally enabled device closest to the data source. A separate distinction shall be made for preprocessing devices—that they are affected by the circumstances of the sensory node. Such a device may not always be present, as specific sensory nodes may e.g. transport their data directly via some form of low power, long-range communication. The more extreme the circumstances of a sensory device, the less performant the preprocessing device can be expected to be due to battery or spatial constraints. As the preprocessing device is often more constrained, it is best suited for simple filtering. This is also more optimal at this level as it loosens network bandwidth and computational power requirements. Significant consideration should be taken regarding the amount of data discarded. However, simple processing algorithms typically yield less accurate results for complex data sets. Furthermore, learning algorithms and neural networks may need *uninteresting* data to build a context for the system's expected behavior. Thus, a good balance needs to be found between filtering that's too wide, resulting in higher resource usage, or too narrow, resulting in false positives/negatives.

1.6.2.3 Edge/fog processing

The edge processing layer can be situated at the first node, where data are obtained from multiple sensory devices. If a preprocessing layer is not available, the filtering is additionally performed here. This will often be the case in resource-constrained environments. In most cases where preprocessing is possible, the device will not be powerful enough to perform all of the tasks mentioned above, or it is not desirable to stress the device.

For quite some time, the idea of edge processing and its extension even closer to the client device—for processing, has been overshadowed by the pure cheap scalability available in cloud services. Therefore in many setups, the edge processing is limited to the aggregation of data. The rise of 5G has led to many discussions of pushing the processing tasks back toward the client. The main driver for these discussions is latency. As many of the over the air transport protocols introduce a significant delay in the processing chain, the additional delay added by transporting the data to the cloud via cable may be considered negligible. However, 5G brings much lower latencies. This leads to ideas of over-the-air applications being used in mission-critical applications such as safety and medical environments. These scenarios where transportation to the cloud may be considered significant give rise to ideas for moving all processing tasks toward the client.

Considering a less critical scenario, however, the primary role associated with edge processing is the aggregation of data. The second task that may be performed here is preparing streams for processing. The data mentioned above enrichment is a possibility for achieving this. Data enrichment is exciting at this level due to the elevated available resources while still being close to the data source. Being closer to the data source may be considered the aspect of the *locality*. Essentially, being physically closer to the source allows for a more accurate labeling of the data and faster processing to achieve lower latency and round trip time. Another good way to prepare the data is to intelligently group and segment streams. This way, a high grade of parallelization can be obtained. For example, consider an edge node that receives data from multiple cars in the surroundings. The stream data can be grouped to create a stream of information for each road, which can then be computed in parallel. But, because this is an aggregated stream, it would require higher computing performance. Yet, the job can be parallelized further by segmenting the stream into multiple time frames. Each segment can then be processed in parallel as well. Such a high grade of parallelization is possible in an event processing context, as events are independent of each other and can thus be identified, extracted from the stream, and processed.

1.6.2.4 The cloud computing

This is where the deep data analysis step will usually occur. Utilizing the performance available in these types of architecture, more complex processing may occur. Typically, more intelligent data processing (e.g., machine learning) is applied due to their higher processing and storage requirements. This is also the main location for persisting the data in storage for further offline analysis.

Every project starts small

What type of processing platform may or may not be more desirable is heavily influenced by the application context. Major driving factors for deciding on a processing platform include the ease of adaptation, performance, scalability, licensing costs, suitability of the architectural model, etc. Certain critical aspects of an application may heavily impede the ability to adopt specific solutions. Many applications will rely on customized solutions for various reasons related to the former driving factors. An application may have started small and hands-on and expanded without refactoring the global design until a fully customized application is used. Such customized applications may not be optimized for big data processing, but the main goal remains to get the job done. Similarly, a decision may be made for a more hands-on approach to lower the implementation time when a project is urgent or needs to interface with an existing external, older interface.

The end goal, however, is to have a system that gets the job done. In many cases, these hands-on solutions will end up reaching a ceiling of their capacities. It will be lower than the case if an existing dedicated stream processing platform were used, or the cost for expanding and maintaining the system reaches a point where the current, costly decision is made to be a refactor and adopt more dedicated solutions. These solutions, decidedly aimed at processing significant data streams, offer efficiency, scalability, flexibility, and failover, but typically will take longer for a company to adapt to their existing solution. A summary of some of these dedicated processing platforms is given.

Open source platforms The apache organization is the leading go-to for opening many open source libraries. This includes several streaming infrastructure components. These are

Kafka Apache Kafka serves as an optimized subscription system for stream processing purposes. The publish-subscribe system is ideally suited for intelligently partitioning streams, which allows for a more optimized parallel processing. Another benefit of this grouping of data by topics is that it allows data to be written away in larger blocks as the data belongs together. Larger sequential writes and contiguous memory blocks offer higher performance for the system. Kafka also offers a stream processing API compatible with the apache stream processing platforms such as Flink and Spark.

Spark Apache Spark relies on what is referred to as RDD (resilient distributed data-set). It is a fault-tolerant, restricted form of distributed shared memory. An API is built on top of this distributed shared memory that enables users to insert various languages into it. This simplifies the implementation of a distributed computing system for developers while making optimal use of clustering technologies. Internally when processing streams, spark uses mini-batches to process the data in the distributed memory. Due to the design model, spark requires some form of clustering infrastructure such as Hadoop YARN or apache-spark native cluster and a distributed file system technology, such as Cassandra or Hadoop distributed file system.

Flink Flink is similar to spark in terms of what it brings to the table. Like a spark, it offers a high-level API for processing streams. However, the processing paradigm is slightly different, where Flink performance processing over the actual streaming data rather than using a distributed shared memory approach. This is done by parsing a program into a DataFlow graph and deploying the various jobs across nodes in the cluster using a job manager master node. Flink does not require a distributed file system but may use various types of storage as data sources or sinks, including compatibility with Kafka. The streaming approach used by Flink allows it to evaluate without batch processing, thus obtaining more responsive processing times.

Commercial platforms The commercial cloud computing industry expands its support for IoT development by creating its own stream processing platforms and creating entire ecosystems that support building IoT applications. Aside from hardware availability, there's additional support in maintenance and system overview, resource management, software APIs for various recurring IoT paradigms (e.g., machine learning libraries, message queue systems, etc.), ease of access via VPN or proxy, etc. Many of the IT giants have their ecosystems. Some examples are IBM IoT Watson, Microsoft Azure, and Google Cloud Computing.

1.6.2.5 The client

The client holds a unique place in the processing chain as its requirements concerning processing are affected by source and implementation design choices. Many of the application-specific processing requirements may be offloaded to the cloud using server requests. The client may typically be expected to hold a certain degree of computational performance. With the introduction of many smart devices to the IoT—the client and the data sensing level are increasingly more often integrated into the same device. The consequences and possibilities of this are elaborated further in the next section.

1.7 Architecture decisions—The modern IoT client

In IoT systems, data sensing and data processing layers are usually not split on an architecture level. In general, there are two common causes to be found. Either an IoT device is present that is simply capable of sensing and transmitting data, or what can be referred to as an intelligent device is present that is capable of performing some computations. Hence, a distinction can be made between computations performed inside the IoT Layer, IoT sensing and processing, and the consecutive edge processing layer. A decision needs to be made on the same tasks to be performed in each of the layers. A significant factor in this decision making is the computing capacity available at each level. If the IoT layer only has data sensing capacities, no more processing tasks can be performed there. However, more tasks such as filtering, simple statistical analysis, error detection, etc., can be performed here in a smart device.

The main contrast between the two layers is to be found in the expected computational power of the devices. While smart devices' presence and processing capacity have been increasing, it is bound by limitations such as available space and energy consumption. By contrast, an individual computing capacity can be assumed (e.g., one or several Raspberry Pi). The layer can handle processing tasks such as data labeling and data aggregation. These are, by nature, the tasks of the edge computing layer. Labeling of data increases the data size and is hence not desirable at lower levels where data rates are more concerning. Aggregation is the natural task, as the edge was described as the first layer obtaining data from multiple sensory systems. Yet more preprocessing tasks can be added, depending on how "fat" the IoT layer is.

Deciding on a "thin vs fat" IoT layer should consider some other factors than just the available computational power. A thin layer minimizes the used power resources at the IoT level. It also simplifies the model and does not set any hardware requirements. However, moving to a more fat IoT layer allows more data to be filtered through analysis, lowering the necessary network and computational specs at subsequent layers. Furthermore, specific tasks may require a low response time, which is easier to obtain if they are performed close to the sensing layer, as network delays are eliminated. These advantages come at the price of needing more computational power at the IoT level, implying a higher power consumption. In the filtering process, the decision-making becomes more complicated because the filtering accuracy has to be considered. A rough analysis implies that the amount of data passing the filter is too much, requiring more network bandwidth or too little, losing some relevant data. A more accurate analysis, however, again requires more processing power. This results in a three-way trade-off between system specs at later levels, computational overhead in the IoT layer, and processing accuracy of the overall system. A balance has to be found between them depending on the specific system context.

1.8 Data enrichment from datastreams for enhanced reasoning

As the paradigm changes from a web to a streaming context, the objective of performing data enrichment changes. In a web context, enriching a web page using XML annotation and linking a web page to other pages using URLs has massive benefits. Not a minor aspect in this is using a standard for said enrichment, so search engines and other applications may use it. Simply put, this allows a webpage to be visited more frequently. More visitors directly implied more revenue due to the advertisement-based model.

The same ideas do not hold for data streams. Data are valuable, and thus it is no longer desirable to share it with others via URL—instead, the objective of the enrichment changes. Data enrichment adds valuable contextual information that enables reasoning at later stages. As mentioned previously, the notion of time and place/space is critical in stream processing. It can be considered the very minimum of data enrichment needed to be performed in almost all systems.

For example, consider the system's goal of alerting security services when a flagged license plate is found. For this purpose, cameras are installed along the road that processes a video stream to detect license numbers. If the cameras emit the license plate number and this gets forwarded straight to the cloud for processing, it can only alert a flagged number has been detected. The questions of *when* and *where* remain unanswered. Answering said the question would require adding some form of timing and location information to the data.

For this purpose, roadside units (RSU) are installed to aggregate and enrich the data before transmitting. Before transmitting the license plate number observed, the processing unit adds a timestamp and possibly a coordinate or the camera's ID that detected this number. If cameras may not be identified uniquely, the ID of the RSU may be used. A server service may connect these IDs to more detailed information using a database. This would allow the maximum amount of information to be contained, with minimal data overhead.

A key motivator for performing this enrichment at a more early stage is the complexity of obtaining equal information at later stages.

1.9 IoT streaming in Smart City applications: Road quality and safe driving

As a case study, consider the idea of using a car's CAN data to analyze road quality in a city. The main goal being the usage of the data for detecting potholes in the roads. Potholes are problematic as they may damage a car, and the road maintainer may be held accountable. They may also lead to injuries. Timely detection and repair of these holes would allow this to be avoided. This timely detection and repair are especially crucial in winter, when these holes may cause significant frost damage to the roads (Hangbin et al., 2019; Huidrom et al., 2013).

A car's CAN bus data is most interesting in terms of IoT processing. First and foremost, it should be considered that each manufacturer of a car usually uses a unique configuration to identify various components of the CAN bus. This configuration is typically not publicly available, which would make the design of a one-fits-all solution especially difficult. Although the idea of reverse engineering this configuration and its related process are most interesting, it will be ignored for continuation.

Next, a CAN bus is a shared data frame connecting the many kinds of sensors in a car. A CAN frame uses a detection mechanism that sensors use to see if they are currently allowed to publish data. This implies the stream is highly heterogeneous. It also implies a certain unpredictability regarding the published data. The sensors may produce information at a specific rate. Still, due to the detection mechanism and possible conflicts, the type of data of the following message is hard to predict and needs to be identified uniquely.

Examining the defining generation rate, the data rates on CAN busses vary. The peak data rate of a realistic scenario is up to 1 Mbit/s. They are considering the available long-distance communication technologies. A prime candidate for transmitting data in this scenario would be any cellular technologies, i.e., the 3G/4G/5G standards, whichever is available, as there is no immediate reason to assume severe power constraints. Experiments show a realistic data rate of 130 Kbit/s using the 3G network at 30 Mph, so at the very least, 4G coverage is a requirement. This would already make the application impossible within certain remote regions due to a lack of coverage. One might consider a type of solution temporarily persisting the data in memory or storage in the car until sufficient coverage is restored. Even if 4G coverage is present, however, 1 Mbit while moving would require excellent coverage and may still be hard to reach. Even considering a full 5G coverage, however, which would make such data rates achievable, the cost of consistently transmitting at 1 Mbit/s for extended periods would be huge.

For the scope of the application, however, the events and data of interest are not contained in the complete CAN bus information. Instead, the data of interest may be singled down to the rotational speed of each wheel. When a wheel hits a pothole, this is assumed to trigger a spike in the specific wheel's rotation speed (see figure below for a graphical representation of a sample of the data set).

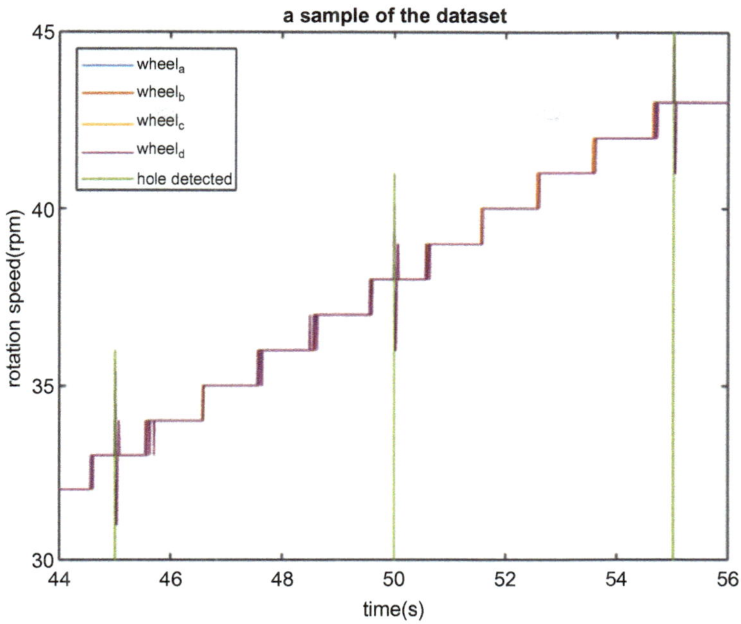

Each sensor uses an identifier on the CAN bus. A mini processing unit could perform a filtering process described previously simply by only retaining the data based on the ID of the rotation sensors. This, of course, relies on the previously stated assumption that these IDs may be known. Such an operation drastically cuts down the system requirements in many ways. For example, the required data transmission rates may be assumed to be divided by multiples of tens. The frequency decreases when the data are transmitted to the server, providing the server with more time for processing. See the figure below to exemplify the arrangements of the layers, from sensing to the end-user client application.

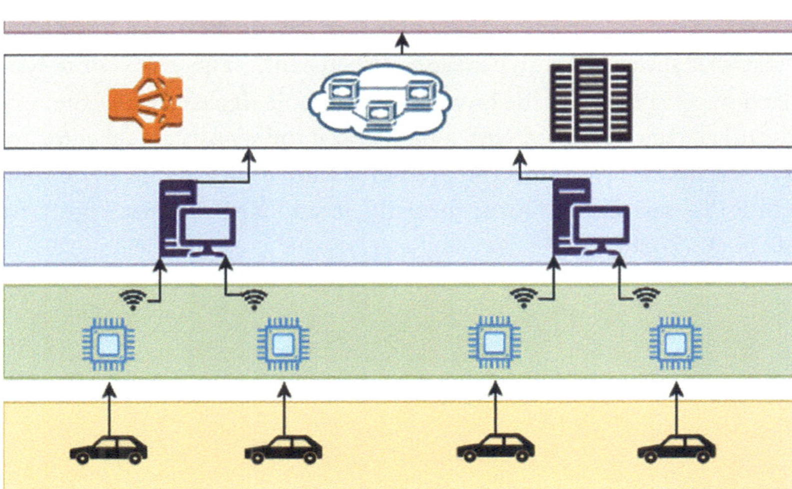

Next, consider the event of interest for this application: detecting potholes inroads. At first glance, this may seem trivial. Whenever a spike in a car wheel's rotation speed is detected, a pothole is found. Real roads, however, have all sorts of possible *obstacles* that may cause spikes that are false detections. E.g., a car may drive over a speedbump, cross a road transition, drive over a sidewalk, etc. Each of these events will result in some sort of spike in the data. Discerning these various spikes and concluding may not be assumed to be achievable using simple algorithms. A typical solution would be to train a machine learning algorithm for this case, as it is well possible to drive a car around and generate a stream that may be used for training. It should also be noted that the goal would be to notify in a timely fashion regarding these potholes. However, the processing itself is allowed to take quite some time as it is less critical than the time needed to reach the hole and repair it.

The scalability of the system depends on how many cars may be equipped with this system. Ideally, as many cars as possible should be equipped to achieve an accurate coverage of all roads. This depends on the degree of modification needed to achieve obtaining and transmission of correct data. In an ideal scenario, a program could be installed on the car's onboard computer for achieving the filtering. The only extra step required would be to plug in a 4G/5G module into the car's available USB slots for transmission. The system could be installed in the cars of public services such as post and police, which would create a system that has well over several thousands of devices involved for mapping a smaller country.

For the enrichment step, relevant context information could be added to the data to improve the knowledge extracted from the data stream. The system should ideally send data in a combined structure that contains some means of identifying the car, its location, and the actual relevant wheel rotation information. Each of these pieces of information should be labeled in a way that allows them to be distinguished easily. This enables the system at a later stage to pinpoint the location of a pothole. It also enables the system to discard all data identified with a single car if the sensors are detected to be faulty. See figure below for a snippet of a sample of semantically enriched data in RPi—serialized in turtle format (left) and XML format (right), from Xhafa et al. (2020).

```
@prefix ns1: <http://schema.org/> .
@prefix ns2: <http://www.w3.org/ns/sosa/> .
@prefix ns3: <http://qudt.org/schema/qudt#> .
@prefix ns4: <http://www.w3.org/ns/ssn/> .
@prefix ns5: <http://www.w3.org/ns/prov#> .
@prefix rdf: <http://www.w3.org/1999/02/22-rdf-syntax-ns#> .
@prefix rdfs: <http://www.w3.org/2000/01/rdf-schema#> .
@prefix xml: <http://www.w3.org/XML/1998/namespace> .
@prefix xsd: <http://www.w3.org/2001/XMLSchema#> .

<edgeprocessor/27657/data/0> a rdf:Bag ;
    rdf:li <car/24898/measurement/0/value/0>,
        <car/24898/measurement/0/value/1>,
        <car/24898/measurement/0/value/2>,
        <car/24898/measurement/0/value/3> .

<car/24898/measurement/0/value/0> a ns1:Observation ;
    rdf:Property [ a ns2:PropertyValue ;
            ns2:name "AnomalyValue" ;
            ns2:value 7.3e-01 ] ;
    ns5:generatedAtTime "2.09" ;
    ns2:hasFeatureOfInterest <car/24898/wheel/0> ;
    ns2:hasResult [ ns3:Unit <http://qudt.org/vocab/unit#RevolutionPerMinute> ;
            ns3:numericValue 0e+00 ] ;
    ns2:madeBySensor <car/24898/wheel/0/rpmsensor> ;
    ns2:observedProperty <car/24898/wheel/0/speed> .

<car/24898/measurement/0/value/1> a ns1:Observation ;
    rdf:Property [ a ns2:PropertyValue ;
            ns2:name "AnomalyValue" ;
            ns2:value 1.6e-01 ] ;
    ns5:generatedAtTime "2.09" ;
    ns2:hasFeatureOfInterest <car/24898/wheel/1> ;
    ns2:hasResult [ ns3:Unit <http://qudt.org/vocab/unit#RevolutionPerMinute> ;
            ns3:numericValue 0e+00 ] ;
    ns2:madeBySensor <car/24898/wheel/1/rpmsensor> ;
    ns2:observedProperty <car/24898/wheel/1/speed> .

<car/24898/measurement/0/value/2> a ns1:Observation ;
```

```xml
<?xml version="1.0" encoding="UTF-8"?>
<rdf:RDF
   xmlns:ns1="http://schema.org/"
   xmlns:ns2="http://www.w3.org/ns/sosa/"
   xmlns:ns3="http://www.w3.org/ns/ssn/"
   xmlns:ns4="http://www.w3.org/ns/prov#"
   xmlns:ns5="http://qudt.org/schema/qudt#"
   xmlns:rdf="http://www.w3.org/1999/02/22-rdf-syntax-ns#"
   xmlns:rdfs="http://www.w3.org/2000/01/rdf-schema#"
>
  <rdf:Description rdf:nodeID="N98d88ed698f845d4af6607e69add8948">
    <ns1:value rdf:datatype="http://www.w3.org/2001/XMLSchema#double">0.2</ns1:value>
    <rdf:type rdf:resource="http://schema.org/PropertyValue"/>
    <ns1:name>AnomalyValue</ns1:name>
  </rdf:Description>
  <rdf:Description rdf:about="car/12144/measurement/0/value/1">
    <ns2:hasResult rdf:nodeID="result1"/>
    <ns2:madeBySensor rdf:resource="car/12144/wheel/1/rpmsensor"/>
    <rdf:Property rdf:nodeID="N98d88ed698f845d4af6607e69add8948"/>
    <ns2:observedProperty rdf:resource="car/12144/wheel/1/speed"/>
    <rdf:type rdf:resource="http://www.w3.org/ns/sosa/Observation"/>
    <ns2:hasFeatureOfInterest rdf:resource="car/12144/wheel/1"/>
    <ns4:generatedAtTime rdf:datatype="http://www.w3.org/2001/XMLSchema#datetime">
      2019-06-08T18:38:08.957512</ns4:generatedAtTime>
  </rdf:Description>
  <rdf:Description rdf:about="car/12144/wheel/0/speed">
    <rdfs:label>the rotation speed of wheel 0 of car 12144</rdfs:label>
    <rdf:type rdf:resource="http://www.w3.org/ns/sosa/ObservableProperty"/>
  </rdf:Description>
```

References

Ahmad, S., Lavin, A., Purdy, S., & Agha, Z. (2017). Unsupervised real-time anomaly detection for streaming data. *Neurocomputing, 262*, 134–147. https://doi.org/10.1016/j.neucom.2017.04.070.

Alencar, B. M., Rios, R. A., Santana, C., & Prazeres, C. (2020). FoT-stream: A fog platform for data stream analytics in IoT. *Computer Communications, 164*, 77–87. https://doi.org/10.1016/j.comcom.2020.10.001.

Alloghani, M., Alani, M. M., Al-Jumeily, D., Baker, T., Mustafina, J., Hussain, A., & Aljaaf, A. J. (2019). A systematic review on the status and progress of homomorphic encryption technologies. *Journal of Information Security and Applications, 48*. https://doi.org/10.1016/j.jisa.2019.102362.

Benson, N. U., Bassey, D. E., & Palanisami, T. (2021). COVID pollution: Impact of COVID-19 pandemic on global plastic waste footprint. *Heliyon*, 7(2). https://doi.org/10.1016/j.heliyon.2021.e06343.

Calders, T., Dexters, N., Gillis, J. J. M., & Goethals, B. (2014). Mining frequent item sets in a stream. *Information Systems, 39*, 233–255. https://doi.org/10.1016/j.is.2012.01.005.

Chamikara, M. A. P., Bertok, P., Liu, D., Camtepe, S., & Khalil, I. (2019). An efficient and scalable privacy preserving algorithm for big data and data streams. *Computers & Security, 87*. https://doi.org/10.1016/j.cose.2019.101570.

Hangbin, W., Yao, L., Zeran, X., Li, Y., Ao, X., Chen, Q., Li, Z., & Meng, B. (2019). Road pothole extraction and safety evaluation by integration of point cloud and images derived from mobile mapping sensors. *Advanced Engineering Informatics, 42*. https://doi.org/10.1016/j.aei.2019.100936.

Huidrom, L., Das, L. K., & Sud, S. K. (2013). Method for automated assessment of potholes, cracks and patches from road surface video clips. *Procedia - Social and Behavioral Sciences, 104*, 312–321. https://doi.org/10.1016/j.sbspro.2013.11.124.

Kim, T., & Park, C. H. (2020). Anomaly pattern detection for streaming data. *Expert Systems With Applications, 149*. https://doi.org/10.1016/j.eswa.2020.113252.

Lindemann, B., Maschler, B., Sahlab, N., & Weyrich, M. (2021). A survey on anomaly detection for technical systems using LSTM networks. *Computers in Industry, 131*. https://doi.org/10.1016/j.compind.2021.103498.

Loukides, G., & Gkoulalas-Divanis, A. (2012). Utility-preserving transaction data anonymisation with low information loss. *Expert Systems with Applications, 39*(10), 9764–9777. https://doi.org/10.1016/j.eswa.2012.02.179.

Gomes, M. M., da Rosa Righi, R., da Costa, C. A., & Griebler, D. (2021). Simplifying IoT data stream enrichment and analytics in the edge. *Computers & Electrical Engineering, 92*. https://doi.org/10.1016/j.compeleceng.2021.107110.

de Assunção, M. D., da Silva Veith, A., & Buyya, R. (2018). Distributed data stream processing and edge computing: A survey on resource elasticity and future directions. *Journal of Network and Computer Applications, 103*, 1–17. https://doi.org/10.1016/j.jnca.2017.12.001.

Nikas, A., Alepis, E., & Patsakis, C. (2018). I know what you streamed last night: On the security and privacy of streaming. *Digital Investigation, 25*, 78–89. https://doi.org/10.1016/j.diin.2018.03.004.

Ramírez-Gallego, S., Krawczyk, B., García, S., Woźniak, M., & Herrera, F. (2017). A survey on data preprocessing for data stream mining: Current status and future directions. *Neurocomputing*, 0925-2312. *239*, 39–57. https://doi.org/10.1016/j.neucom.2017.01.078.

Rhahla, M., Allegue, S., & Abdellatif, T. (2021). Guidelines for GDPR compliance in Big Data systems. *Journal of Information Security and Applications, 61*. https://doi.org/10.1016/j.jisa.2021.102896.

Sopaoglu, U., & Abul, O. (2020). A utility based approach for data stream anonymisation. *Journal of Intelligent Information System, 54*, 605–631. https://doi.org/10.1007/s10844-019-00577-6.

Stephanie, V., Chamikara, M. A. P., Khalil, I., & Atiquzzaman, M. (2021). Privacy-preserving location data stream clustering on mobile edge computing and cloud. *Information Systems*. https://doi.org/10.1016/j.is.2021.101728. In press.

Xhafa, F., Kilic, B., & Krause, P. (2020). Evaluation of IoT stream processing at edge computing layer for semantic data enrichment. *Future Generation Computer Systems, 105*, 730–736. https://doi.org/10.1016/j.future.2019.12.031.

Zhou, Y., Ren, H., Li, Z., & Pedrycz, W. (2021). An anomaly detection framework for time series data: An interval-based approach. *Knowledge-Based Systems, 228C*. https://doi.org/10.1016/j.knosys.2021.107153.

CHAPTER TWO

Stream processing in the semantic web

In this chapter, the concept of semantic data streams and semantic stream processing are discussed as topics of interest in the context of smart cities and collective intelligence. The discussion is done from the perspective of the semantic web. In order to do so, the semantic web technology basis is first elaborated. The RDF-Stream format is discussed as a means of representing semantic data stream, with TripleWave as a possible means of generating these streams. The semantic web technologies described here can be found useful in the processing of semantic IoT data streams by reusing the advantages of semantic web technologies in data stream environments. Some challenges and barriers in reasoning over RDF streams are discussed. The reader is referred to this chapter of this book for models and properties of IoT data streams.

2.1 Semantic data streams

2.1.1 Basic concept

The first issue in the process of designing an independent, event-driven IoT architecture is the notion of meaning. At the edge of a system, raw data is generated by sensor nodes. Due to its heterogeneous nature, problems arise with relation to interoperability and interpretation. In order to be able to define events, it is first necessary to have interpretable data. To make data machine-interpretable, so-called metadata is added. Metadata is data that provides additional information about data, based on preexisting knowledge. This metadata is often described using markup languages, e.g., XML and JSON. This is also referred to as semantic markup.

The addition of semantic markup allows the definition of attributes through domain-specific vocabulary tags. This improves the specification of what kind of data is being passed, but still leaves a few issues. The vocabulary used for the tags has to be agreed on, which results in many domain-specific standards for the metadata. Furthermore, semantic markup only

Collective Intelligence for Smart Cities
https://doi.org/10.1016/B978-0-12-820139-8.00005-X

defines *what* the data is about. It does not enable the description of *how* it is related to other elements. The latter issue was already discovered and tackled in the context of web technologies, and led to the establishment of the various semantic web standards, set by the W3C.

New challenges were then introduced by the IoT age and adaptations were made, aiming to reuse the advantages of the semantic web technologies in data stream environments. Indeed, IoT data streams introduce various characteristics (data volume, velocity, heterogeneity, etc.) which are not present in the standard semantic web setting.

2.1.2 Semantic web

The semantic web is an extension of the original web, with the twofold goal of both linking as much data as possible and giving a well-defined meaning to the data, such that it is more suitable for both computers and people to cooperate. This resulted in the creation of the semantic web stack (Fig. 2.1).

The basis of the stack shows that everything is built upon the existence of web-unique identifiers (URI and IRI). The Resource Description Framework (RDF) uses these identifiers to describe relationships between

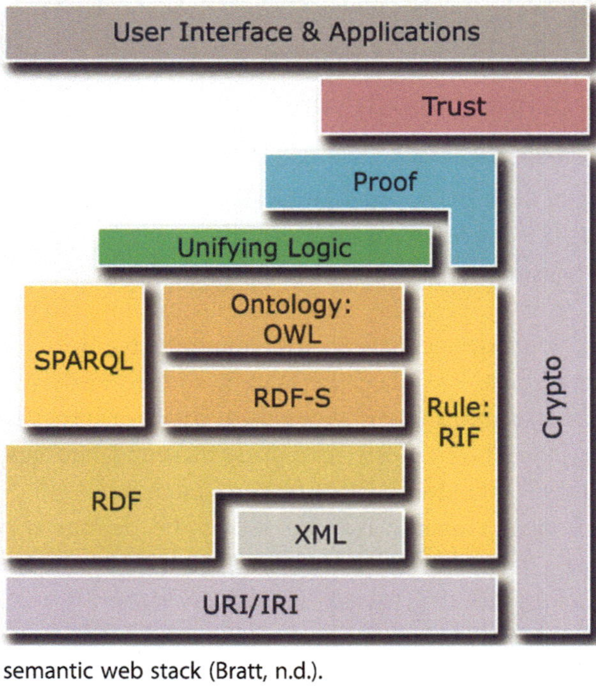

Fig. 2.1 The semantic web stack (Bratt, n.d.).

resources on the Internet. This is done through statements in the form of so-called "triples" that have a *subject-predicate-object* structure. Originally RDF was suggested as a standard to be built upon the XML markup language, but later it was extended toward other formats as well (Raimond & Schreiber, 2014).

The basic RDF specification allows for the definition of relationships between objects, but it does not contain a distinct ontology for describing these relationships. A basic ontology that is focused on describing the structure of data, i.e., the hierarchy, was added in the first extension called RDF-Schema (RDF-S) (Brickley & Guha, 2014). A far more extensive ontology is included in the Web ontology language (OWL). The OWL standard includes semantics for logical expressions, e.g., equality, restrictions, and cardinality (Parsia, Hitzler, et al., 2012). Another key difference between OWL and RDF-S is that OWL also introduces the concept of invalid usage of the vocabulary. The extended vocabulary in OWL allows contradictions to be detected. An ontology with contradictory definitions is considered inconsistent (Parsia, Motik, & Patel-Schneider, 2012).

The W3C acknowledged that no single one-fits-all rule language could possibly satisfy the needs of different paradigms used in knowledge representation and business modeling (Boley & Kifer, 2013). Hence, rather than developing a standard for rule systems, the Rule Interchange Format (RIF) was designed to facilitate the interchange between different rule engines. The goal of RIF is to enable rule sets to be communicated from one system to another, using a dialect they both support. This requires the rule engines to have a semantics-preserving mapping to the RIF dialect (Boley & Kifer, 2013).

Finally, SPARQL provides a means of querying RDF content. At the time of writing, there are no existing standard specifications for the trust, proof, and security layers in the semantic web stack. For security reasons, the usage of a digital signature is recommended.

2.1.3 RDF-stream

The transition from a web to a streaming context introduces a number of issues for the semantic model. The RDF Stream Processing (RSP) Community group was created to find ways to define a common model for handling RDF Streams and to tackle these issues. Their research resulted in the proposal of an extension to the RDF standard called RDF-Stream, which in particular focused on handling the aspect of time. At the time of writing, it has not been accepted as an official W3C specification and is marked as incomplete (Anicic et al., n.d.), even though it has been widely adopted.

In the RDF-Stream draft (Anicic et al., n.d.) a proposal is made to use time annotations. This can be done in two ways: using a description of data as a point in time, or as existing within a specified interval. Three types of metadata are recommended to use in this case:

(1) In case of a single point in time annotation, the use of either the *production time* (time when the data element was produced, if possible) or the *receiving time* (time the data element is made available to the RSP) is recommended.

(2) In case of an interval, the annotations *start time* and *end time* can be used to specify the validity of the data.

(3) A *timestamp predicate* should be used for a timestamp presence. The use of *prov:generatedAtTime* or *ssn:observationSamplingTime* is suggested.

It should be noted that in a lot of cases, this time specification does not have to be an actual time value. Specifying the order of the elements in an incremental way is sufficient to allow processing of the data. Furthermore, the possibility of using the order of arrival at the RSP as an implicit timestamp is mentioned for RDF streams that do not have time annotations upon generation (Anicic et al., n.d.).

An RDF-Stream should be identified with an internationalized resource identifier (IRI). There are no further requirements set regarding this IRI. There is no specific indicator that an IRI is an RDF stream.

As a final note, it is worth mentioning that an RDF-Stream is nothing more than a stream of RDF graphs that are required to have at least a time annotation. The timestamps are added using standard RDF triple definitions. Hence, RDF-Stream still is a subset of the original RDF standard and is fully compliant with it.

2.1.4 TripleWave

TripleWave is an open source tool created in collaboration by Politecnico di Milano and Ecole Polytechnique Fédérale de Lausanne. Its goal is to provide a means of publicizing RDF streams on the web. It is provided as Javascript and can convert web streams to RDF-streams. In order to do so, it uses the RDB to RDF Mapping Language (R2RML) (Sundara et al., 2012). It also provides support for replaying an RDF dataset, which can provide useful applications such as benchmarking, system evaluation, and simulation.

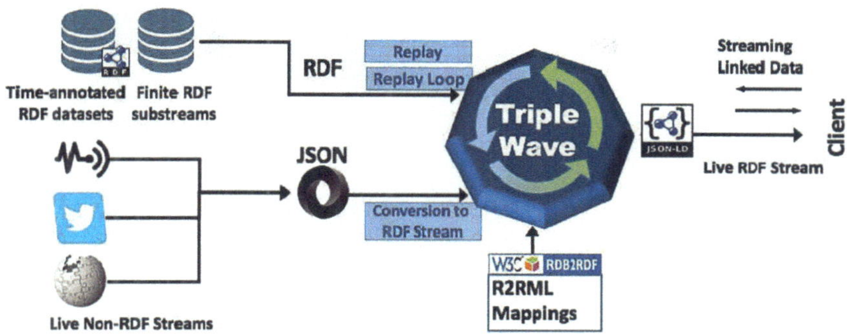

Fig. 2.2 The architecture of triple wave, for generating RDF-Streams (Mauri et al., 2016).

Furthermore, it is also possible to loop over these datasets in order to simulate an infinite data stream (Fig. 2.2).

It is important to notice that the TripleWave framework runs on top of JSON formatted input and generates a JSON-LD RDF stream as output. In order to satisfy this assumption, additional conversions may be needed to match input data for the TripleWave processing, or output data for the stream processor.

2.2 Semantic stream processing

Stream processing systems have the goal of continuously evaluating incoming data in order to find certain data of interest. This data of interest has to be found real time (generally within a few seconds) or near real time (e.g., within a few minutes). This is in contrast to batch operations, where the data is stored in a storage layer and then queried for processing at a later time and can be accessed as many times as needed. One particular data pattern of interest is an *event*. The most primitive of events are created by evaluating basic relational operators (equality, more or less than, …) over a single data attribute at a single point in time. Multiple of these events can be combined to create composite or aggregate events.

The relationship between the description of these types of events and the querying of a relational database is obvious. If imagined as taking a snapshot of a stream at a single point in time, one could save and perform a relational database query upon the data to find these kinds of events. This approach to reasoning with stream data is called data stream management systems (DSMS) (Cugola & Margara, 2012). The shared characteristic of the systems designed using this approach is that they implement windows to determine which data elements are relevant. These elements could be considered a part of the "snapshot." Time window processing will either accept elements that have a time specification within a certain interval or count a specific amount of elements. The usage of these time windows encourages a single point in time annotation of the data.

The advantage of these types of implementations is that they can build on the established knowledge regarding relational databases. The disadvantage is that the DSMS approach makes it hard to detect spatial events. These are patterns in time, such as sequences, the repetition of a certain event or element, or even the absence of such a pattern (anomaly detection). In order to tackle these kinds of events, the complex event processing (CEP) approach was created. These families of processors have distinct semantics for defining complex events, given the idea of time annotated data and using the principles of primitive and aggregate events. There are multiple approaches to this processing paradigm: using composition operators, production rules, timed finite state machines, and logic languages (Helmer et al., 2011). The *fire* event is a commonly used example, consisting of the detection of smoke followed by the detection of a high temperature within a short time frame.

In the following sections, a handful of the various engines that specifically implement these techniques for reasoning with semantic stream data are provided, starting with the DSMS-based approaches, followed by the CEP approaches. The basis for them is the SPARQL language used in the semantic web.

2.2.1 DSMS approaches

2.2.1.1 C-SPARQL

Continuous-SPARQL (C-SPARQL) (Barbieri et al., 2010) is a language for continuous queries over RDF data. It extends the SPARQL 1.1 standard, used for querying RDF data in the semantic web. In order to achieve this, a *register* query statement is added that causes the registered query to be run at frequency specified by a *computed every* clause. The register query also specifies the format of the output—as a new RDF stream or as a query result, possibly in the form of RDF graphs or tables of variable bindings. In the case of an RDF stream output, an additional *construct* statement must be added to specify the generated triples. The timestamp of the generated RDF triples will match that of the query execution time, regardless of what the timestamp of the involved triples is. An RDF stream output can be reused for processing in other queries.

In order to execute these stream operations, C-SPARQL assumes that each RDF triple in the stream contains a timestamp that does not have to be unique, but must be increasing. In case multiple timestamps have been bound, the most recent timestamp of the bindings is taken. The absence of a timestamp results in a type error. A time window is then used to query the relevant data from the stream. C-SPARQL implements both a tumbling window and a sliding window.

C-SPARQL does not implement specific operators for performing the detection of temporal events, but does provide a function to query the timestamp of stream data elements. Hence it is still relatively simple to construct queries for doing so by using this timestamp. In this way, C-SPARQL supports the processing of complex events to a certain degree. A restriction in this regard is the fact that all the relevant data for a temporal event has to be contained within the same window of the query (Margara et al., 2014). A bigger window requires more memory, so the detection of temporal events that are spread wide in time is troublesome.

The architecture of C-SPARQL uses a modular approach. In order to handle the continuous queries, a query translator is used to integrate existing technologies. A continuous engine (using both Esper and Jena) is used to determine relevant triples from the stream. The outcome of this process generates a snapshot of the data, which is processed by a regular SPARQL engine.

2.2.1.2 SPARQLstream

Another approach based on the delegation of continuous queries to existing engines is proposed in SPARQLstream (Calbimonte et al., 2010). While similarities with C-SPARQL are clear, it should be noted that SPARQLstream is in fact very different in execution. C-SPARQL uses a delegated engine to execute the window operators to create a snapshot of the relevant data, but the processor still processes RDF data using a SPARQL engine. In the SPARQLstream approach, this is no longer the case. The complete processing of the data is done using existing relational database stream engines.

In SPARQLstream the incoming RDF-streams are translated to relational data using the R2RML language. In order to process the querying of the data, a morph-stream engine is used. The morph-stream engine transforms the query in a two-step process. The first step of the process is to translate the query into relational algebra expressions that have been extended with window constructs. In the case of SPARQLstream, this window is exclusively time based. This intermediate step brings two advantages:

- Logical optimizations can be applied (by applying join and union distribution, and pushing down projections and selections)
- More flexibility regarding the used relational stream engine, e.g., translations to SNEE and ESPER (Calbimonte et al., 2012).

This algebraic form is then translated into the respective language of the engine. This implies that the limitations of the target language have to be taken into account as they may prevent a complete translation, e.g., in SNEE statements for sequences and disjunction are not supported (W3, n.d.).

The final step in the process is to translate the resulting tuples into RDF triples. When this inverse mapping happens, the resulting RDF-Stream is marked as a *virtual RDF-Stream*. Its IRI is accordingly described in a virtual stream element. This annotation is made since the stream is derived from the data resource, unlike RDF-Streams that are created as a result of a query on RDF data in C-SPARQL.

While the SPARQLstream approach no longer relies on a SPARQL engine, its query semantics still are based on an extension of the SPARQL1.1 standard.

2.2.1.3 CQELS

The Continuous Query Evaluation over Linked Streams (CQELS) language[a] attempted to answer possible performance issues that occur by using the "black box" approach, which uses translations to existing processing engines for dealing with stream processing (Le-Phuoc et al., 2011). In order to attempt to further improve better query optimization, an engine was designed that implements the necessary operations for dealing with streams. More precisely, the CQELS engine implements:

- Window operators: Used for extracting triples from an RDF stream, given a triple pattern and a time window.
- Relational operators: Traditional relational database operators, e.g., join and union.
- Stream operators: An operator needed to generate RDF streams from the sets that resulted from the relational operators.

Details on the semantics for these operators can be found in Le-Phuoc et al. (2011).

[a] See for instance GitHub—cqels/CQELS-1.x: CQELS engine forked from Google Code (LGPLv3.0).

The resulting CQELS query language (CQELS-QL) thus extends the SPARQL language.[b] A query pattern for applying window operations on RDF streams is included in the StreamGraphPattern (Listing 1).

StreamGraphPattern ::==	**'STREAM'**[*Window*] *VarOrIRIref*{*TriplesTemplate*}
Window ::==	*Range* | *Triple* | 'NOW' | 'ALL'
Range ::==	'RANGE' *Duration*('SLIDE' *Duration*)?
Triple ::==	'TRIPLES' INTEGER
Duration ::==	(INTEGER 'd' | 'h' | 'm' | 's' | 'ms' | 'ns')

Listing 1 The StreamGraphPattern, in quoted capitals keywords that are used in the query language (Le-Phuoc et al., 2011).

The various keywords are used in the creation of different types of windows. For a time-based window, the *range* keyword must be used to specify its duration, which in turn is specified by an integer value and a unit of time. In addition, the *slide* keyword can be used to specify how fast a sliding window moves, otherwise a tumbling window is assumed. For a triple-based window, the keyword *triples* is used to specify the amount of triples that are kept within the window. The keyword *now* specifies that only triples that have the current timestamp will be kept. The keyword *all* causes all triples to be kept in the window.

2.2.2 CEP approaches

2.2.2.1 EP-SPARQL and ETALIS

EP-SPARQL (Event-Processing SPARQL) is a processing engine that uses a CEP approach. In fact, it was designed with the issue in mind that the other systems at the time all used a DSMS approach and hence struggled to detect particular sequences in the data.

Like the previous implementations, EP-SPARQL is an extension of the original SPARQL language. More specifically, it implements the logical operators *SEQ, EQUALS, OPTIONALSEQ*, and *EQUALSOPTIONAL* (Anicic et al., 2011). The syntax for defining temporal relations using these operators is the following:

- P_1 SEQ P_2 joins if P_2 occurs after P_1
- P_1 EQUALS P_2 join if P_1 and P_2 occur simultaneously.

The *OPTIONALSEQ* and *EQUALSOPTIONAL* operators are time-sensitive variants of SPARQL's *OPTIONAL* query syntax, which allows

[b] See for instance https://github.com/streamreasoning/CSPARQL-engine.

making additional queries that will not cause the overall query to fail if they cannot be met.

EP-SPARQL does not implement a window-based approach, but rather assumes the interval-based RDF-Stream model. In order to reason with these intervals, functions for obtaining the duration, start time, and end time of the data are implemented.

Another important aspect to note is that EP-SPARQL does not implement a specific approach to handle negations. In order to achieve negation, i.e., the detection of the explicit absence of a certain triple pattern, the pattern must first be queried in an optional construct with the result binding to a specific variable, followed by a filter that states that the variable is not bound. Hence the optional pattern has failed to find a match, and thus the absence of a pattern has been detected.

The EP-SPARQL language was developed in order to enable the ETALIS engine to be used in real-time semantic web applications (Anicic et al., 2012). The ETALIS engine is designed to be a pluggable environment for various prolog-rule-based systems. These prolog systems are then used to generate event streams, based on prolog code that is generated by parsing the native ETALIS language for events into binary rules and compiling the binary rules into specific prolog code.

By extension, this system can be used to process EP-SPARQL queries. This simply includes more parsing steps, to transform the EP-SPARQL syntax into the internal prolog format used in the ETALIS engine.

One of the advantages of this approach is that the background knowledge included in the semantic annotations is also parsed to prolog code and hence passed on to the prolog system. The prolog system may then be able to use this to infer implicit information. Another advantage is created by the pluggable nature of the environment. Certain prolog systems may be more suitable depending on the nature of the queries. If it is known that a specific type of query (e.g., aggregation process for environmental data) will be used more often, an appropriate prolog system with better performance can be selected for this.

2.2.2.2 CQELS-EP

An extension to the original CQELS engine has been made in Chu et al. (2017). While the CQELS engine does implement a native, "white box" approach to processing semantic streams, it is incapable of temporal correlation of data. To this end, the aforementioned StreamGraphPattern is modified, by replacing the *TriplesTemplate* component with a *triple sequence pattern*. This consists of multiple triple templates connected by temporal relations (see Listing 2).

```
TrTemp ::= TriplesTemplate
TSP :: = TrTemp | TR '(' TrTemp (','TrTemp)* (', ! '
         TrTemp) (',' TrTemp)* ')' | TR '(' ('; ' TrTemp)
         * (', ! ' TrTemp) ('; ' TrTemp)*TrTemp ') '
TR ::= SEQ | After | Meets | MetBy | Overlap |
       OverlappedBy | During | Contains | Starts |
       StartedBy | Finishes | FinishedBy | Equals
StreamGraphPattern ::= 'STREAM' '[' Window ']'
                       VarOrIRIref '{' TSP '}'
```

Listing 2 The new StreamGraphPattern in CQELS-EP. Various temporal relations are defined.

Other adaptations that are made are with regard to processing these complex events. CQELS assumes a point-in-time data model, hence an event expires when it is no longer part of the window. A correct time window must be found, such that the complex event is always detected. Furthermore, CQELS has an incremental implementation, which discards previous results even when they have not yet expired. Such behavior is undesired, as previous results may be needed in the processing. This leads to possible incorrect output or a need to reprocess a query to re-obtain the previous result. To solve this, caching mechanisms are implemented.

The implementation of the temporal logic is largely based on an earlier adaptation (Chu et al., 2017; Dao-Tran & Le Phuoc, 2015). The CQELS implementation can be found on the github page (Lephuoc, n.d.). This adaptation aims to support CEP, persistent streams, RDFS/OWL reasoning, and websockets, as well as supporting the syntax recommendations made by the RSP group (Anicic et al., n.d.). For a suite of metrics to analyze the performance of C-SPARQL and CQELS RSP engines the reader is referred to Ren (2018) and Ren et al. (2016).

2.2.3 Recent developments

IoT semantic stream processing is an active field, with a lot of research work and ongoing projects. As IoT computing expands new fields of engineering, business, finance, health, etc., new research and technological views are brought into this field to overcome the limitation of conventional tools when dealing with IoT data (Karaagac et al., 2019; Valero et al., 2021; Xhafa et al., 2020). Smart cities and collective intelligence are undoubtedly fields of interest for the application of these new findings and developments. The use of semantic web technologies is a source of inspiration for the Internet of Things domain (Rhayem et al., 2020). Advantages of Edge computing are also being sought to achieve scalability of RDF Engines on IoT Edge Devices (Le-Tuan et al., 2020; Nguyen-Duc et al., 2020).

2.2.4 Summative evaluation on semantic data streams

The various processors are designed to continuously query data from a stream source, in order to find possible data of interest. More development will be needed in order to achieve full-fledged IoT semantic stream processing. The processing engines are all standalone models, which are suited for processing on a single centralized device. Such a centralized, standalone approach does not match with the need for clustering and cloud computing that exists for dealing with the overhead generated by the *Big Data* associated with the IoT. Although some have been extended toward this end (Hoeksema & Kotoulas, 2011), their code base is not publicly available and there is still the problem regarding their expressiveness.

Event processing in an IoT streaming context implies dealing with multivariate data and possibly highly complex patterns. Since the current expressiveness of these engines is mostly limited to simple relational database calls that do not support temporal patterns, they will be insufficient for a large number of scenarios. In the future, it will be interesting to keep track of how these engines evolve toward the integration of complex temporal patterns and perhaps the integration of neural networking technologies for handling these.

2.3 Reasoning over data streams

The detection of events involves a process of selecting data of interest from a dataset or stream. Reasoning takes the next step. The goal is to produce new results from original data. In this regard, a very important concept within the semantic context is that of inference. Inference is the process of drawing a conclusion from observations.

The inference of new knowledge is just one of the possibilities for reasoning over events. Another possibility involves the handling of events. When an event occurs, certain actions may need to be taken. In this context, two groups of reasoning systems can be considered: production rules and ECA rules. Production rules or business rules specify actions to be taken when a certain event occurs in an *on* event *do* actions format. The name production rules are related to the fact that a lot of these systems were created in a database context. Hence the action to be taken is the production of a certain element in the database. ECA rules are similar to production rules, but include a condition that has to be validated before the action is executed.

2.3.1 Inference

The semantic web expands the "meaning" in the available data. The addition of these semantics has the goal of enabling machines to process the data more efficiently, but also more intelligently. One goal in particular is to expand the general knowledge, by inferring implicit information from explicit data. The inference process is often referred to as *reasoning* in the semantic context, and the RDF and OWL standards help in doing so. However, because of the extensiveness of OWL's semantics, the inference of new information is often a very computationally expensive process.

In some cases, the reasoning process can even become indecisive (Hoeksema & Kotoulas, 2011). A simple example taken from the W3C OWL reasoning examples (Bechhofer, n.d.) shows how the inference process could work (see Listing 3).

```
Class(a:bus_driver complete intersectionOf(a:person
        restriction(a:drives someValuesFrom (a:bus))))

Class(a:driver complete intersectionOf(a:person
        restriction(a:drives someValuesFrom (a:vehicle))))

Class(a:bus partial a:vehicle)
```

Listing 3 An example of class inference possibilities in OWL.

The OWL statements define a class *bus driver* to be someone who drives a bus. Furthermore, a driver is someone driving a vehicle and a bus is a subclass of the vehicle class. Hence the engine will infer that a bus driver is actually a subclass of the driver class.

In a similar way, the inference process can also be used to find inconsistencies in the data. Such inconsistencies can be for example multiple class definitions of the same object, or contradictory relations, e.g., X is a child of Y followed by a definition Y is a child of X.

While the concept of inference is in fact one of the important aspects when discussing semantic technologies, the scope of this project is limited to reactively handling events in a semantic context.

2.3.2 Production rules

Production rule systems are systems that use rules to solve a certain problem, mimicking the reasoning process of a human expert. Hence they are also known as production systems or expert systems. The basis of the system are rules, which are composed in *if-then* statements. The *if* clause is called the premise or condition, and the then part is called the consequent, conclusion, or action (Grosan & Abraham, 2011). As can be derived from the various terms, rules can roughly be used for two different purposes, one being the aforementioned inference process.

The other is to aid in the decision-making process of systems. A production rule system can be used to perform various tasks such as classification, monitoring, and prediction (Grosan & Abraham, 2011). Important in this regard is the size of the rule set, which greatly influences the performance. In addition, if more rules apply to the same given fact, a conflict resolving strategy has to be applied.

2.3.2.1 Drools

JBoss Drools (Official drools website. Red Hat Inc., n.d.) is an open-source business rule management system. It provides an extensive framework for setting up rule-based systems. It consists of several major parts:

- Drools Workbench: A web user interface for managing, authoring, and testing rules.
- Drools Expert: The business rules engine.
- Drools Fusion: Adds complex event processing features to the system.
- jBPM: A modeling tool for managing workflows. It is meant to bridge the gap between software developers and business analysts. By using flow charts and domain-specific representations, the process becomes more easy to understand by business users. It also increases the system's overall flexibility.
- Optaplanner: A constraint resolving engine for determining efficient paths. It solves scheduling puzzles.

It is interesting to note here that drools combine both the idea of business management and event processing. The choice to combine these two fields was made because of their similarities in requirements and goals. From a business rule point of view, rules are often defined based on the occurrence of certain events, effectively using them to perform a kind of event processing. Furthermore, both events and business rules need to adapt to rapidly changing environments (Drools fusion documentation, n.d.).

Other similarities can be found both with regard to their functional and nonfunctional requirements. Business rules often need to perform pattern matching tasks and are often tied to processing time constraints (Drools fusion documentation, n.d.).

2.3.2.2 *Jena*

Jena (Apache jena, n.d.) is an open-source framework specifically created for building semantic web applications. The Jena framework allows for reading, writing, querying and storage of RDF data. Furthermore, it has components for setting up a SPARQL end-point to expose data on the net.

While the Jena framework is not designed as a production rule system, it does include various rule engines for inferencing over semantic web data. Because of its specific field, it has to be considered as an option when discussing the possibilities for the rule-based, stream processing of semantic data events. In this regard, a study was done to compare the performance and possibilities of Jena and Drools in an event processing context (Apache jena, n.d.).

One of the important conclusions is that Jena is unsuited for performing an inference task over stream data, due to the way the model works. Furthermore, although the inferencing process itself outperforms that of Drools when using semantic data, it is compensated by the time needed to insert the data into the model. Hence Drools is better suited for stream reasoning unless it is necessary to be able to query the data or memory constraints apply (Fobel & Subramanian, 2016).

2.3.2.3 *Easy-rules*

Where Drools provides an extensive framework for creating a business rule management system, Easy-Rules (Hassine, n.d.) is designed to provide a simple framework for rule-based processing in Java. The lightweight API allows users to define rules in various ways: using annotations, in an expressive way, or even using a rule descriptor in YAML format. Due to its simplicity, Easy-Rules allows for fast design of a rule system, but it is not concerned with issues such as conflict resolving, planning, or process design which are integrated with Drools. It does not offer support for processing data in a streaming fashion either.

2.3.3 Event-condition-action (ECA) rules

ECA rules, originally proposed for active databases (McCarthy & Dayal, 1989), are often used as a basis for reactively handling changes in event-based systems. An ECA rule contains a specification of the *event*, which is a certain object, pattern or signal of interest. When such an event occurs, the engine will evaluate the *condition* specified. When the condition evaluates to true, the action that is specified in the *action* part of the rule is executed. In a CEP environment, these rules can be used to handle an occurrence of an event of interest. Such an event could be, e.g., the detection of a hole in the road, for which the position has to be added to the database.

Various extensions to this model have been proposed depending on the application context. One of the possible extensions is the realization that often when for an event certain conditions are not met, other actions may need to be taken depending on further conditions. In this regard, two extensions were suggested (Adaikkalavan & Chakravarthy, 2011), both combined resulting in an extended ECA pattern (see Listing 4).

```
RULE [ R_name
       EVENT       E_name
       CONDITION   C_1 ... C_k
       ACTION      A_1 ... A_n
       ALT ACTION  AA_1 ... AA_n        ]
```

Listing 4 The ECA rule pattern (Adaikkalavan & Chakravarthy, 2011).

The first extension is the suggestion to consider an entire list of possible conditions for a certain event and actions that are associated with it. This way there are less rules necessary to handle each event, making conflicts less likely. The second extension then is the realization that often, when conditions are not met, there is an action that should be taken exactly because an event occurred and its conditions were not met. A simple example for this case could be a web request without the necessary credentials. The server does not execute the request, because the condition of having the credentials is not met. However, it is still necessary to notify a user why his request was denied. Combining these two ideas resulted in the extended ECA, event condition action alternative pattern. The specified alternative actions are to be executed in case the condition does not evaluate to be true.

Some other adaptations to tackle possible issues include the ideas of ECA-P, ECAP, and ECAS patterns (Hu et al., 2017). In the ECA-P (ECA-Postcondition) pattern, postconditions are added in order to avoid conflicting rules. The ECAP (ECA Parameters) pattern adds parameters to the ECA concept. These parameters are then used in the action process. By doing so the amount of data passed to the calculation engine can be lowered in object-driven environments. Lastly, in ECAS (ECA sequence), the idea of using the sequence of previously taken actions in order to detect anomalies is introduced. The common factor among all these adaptations is the need for events to be handled instantaneously, and the usage of events as triggers to start the actions.

2.3.4 Recent developments

Inference and reasoning over IoT data streams have become a building block in all IoT-based applications. Clearly, semantic IoT plays an important role here to enable advanced inference and reasoning systems by using intelligent and machine learning methods. This is a very active field of research and development, whereby new applications and machine learning models are presented and show both the need and strength of IoT semantic reasoning in a variety of application domains, including smart cities and collective intelligence (Chen et al., 2020; Kayan et al., 2021; Le-Phuoc & Eiter, 2020; Lim & Rahmani, 2020; Luo et al., 2021; Tanque, 2021).

The RDF Stream Processing (RSP) community recognizes, however, that the various models and languages proposed over the past decade for continuously querying and reasoning over RDF streams with their semantics poses challenges and makes them hard to compare. A recent advance toward unification is the RSPQL, which provides programming abstractions based on RSPQL semantics and thus enables comparable studies. Another proposal in this direction is RSP4J (Tommasini et al., 2021), a flexible API for the development of RSP engines and applications under RSPQL semantics.

2.3.5 Summative evaluation on inference and reasoning

While the aspect of inference is important in the semantic web context, more efforts are needed to find ways to achieve inference over streaming data in an IoT context, due to the computational expense of the process.

Production rule systems and streaming contexts, however, are greatly complementary, especially for a task such as the classification of data. Push-based semantics are desirable in a streaming context where the reaction time of the system is crucial. In this regard, the integration of complex event processing and streaming features into the Drools rule management system is most promising. In many cases, however, there is no need for such an extensive rule system.

The ECA pattern and its various adaptations all bring a similar, reactive processing style. Rather than having extensive frameworks, the majority of these ECA-based systems are created using simple scripts that use declarative programming languages. Such a design choice can be justified to a certain degree, because in the majority of these examples the reasoning layer is both the final step in the processing chain and performs a simple classification task. This results in small scripts for performing these tasks, where the expressiveness of the used declarative programming languages allows them to be adapted relatively easily if needed.

Despite advances in the field, the RSP engines have to deal with volume and velocity characteristics as well as dynamically changing data structure and workload characteristics of the Big Data streams. While there have been proposed various scalable solutions, namely, adaptive distributed RDF Stream Processing, reasoning over RDF streams, performance, and reliability remain challenging issues.

In particular, such challenges should be addressed in the context of smart city applications based on data streams, where semantic stream reasoning is deemed crucial to support decision processes in smart cities (D'Aniello et al., 2018).

References

Adaikkalavan, R., & Chakravarthy, S. (2011). *Generalization of events and rules to support advanced applications* (pp. 173–193). Springer. https://doi.org/10.1007/978-3-642-19724-6_8.

Anicic, D., Fodor, P., Rudolph, S., & Stojanovic, N. (2011). Ep-sparql: A unified language for event processing and stream reasoning. In *Proceedings of the 20th international conference on World Wide Web, ser. WWW '11* (pp. 635–644). New York, NY, USA: ACM. http://doi.acm.org/10.1145/1963405.1963495.

Anicic, D., Rudolph, S., Fodor, P., & Stojanovic, N. (2012). Stream reasoning and complex event processing in etalis. *Semantic Web, 3*(4), 397–407.

D. Anicic et al. RDF stream processing: Requirements and design principles. RDF Stream Processing Community Group. http://streamreasoning.github.io/RSP-QL/RSP Requirements Design Document/.

Apache jena. *The Apache software foundation*. https://jena.apache.org/.

Barbieri, D. F., Braga, D., Ceri, S., Valle, E. D., & Grossniklaus, M. (2010). Querying rdf streams with c-sparql. *ACM SIGMOD Record, 39*(1), 20–26.

S. Bechhofer. n.d. Owl reasoning examples. http://owl.man.ac.uk/23/why/latest.

Boley, H., & Kifer, M. (2013). *RIF overview* (2nd ed.). W3C. W3C Note http://www.w3.org/TR/2013/NOTE-rif-overview-20130205/.

S. Bratt. n.d. (27) Semantic web, and other technologies to watch. W3C. https://www.w3.org/27/Talks/0130-sb-W3CTechSemWeb/#(24).

Brickley, D., & Guha, R. (2014). *RDF schema 1.1, W3C, W3C Recommendation*. http://www.w3.org/TR/2014/REC-rdf-schema-20140225/.

Calbimonte, J.-P., Corcho, O., & Gray, A. J. G. (2010). Enabling ontology-based access to streaming data sources. In P. F. Patel-Schneider, Y. Pan, P. Hitzler, P. Mika, L. Zhang, J. Z. Pan, I. Horrocks, & B. Glimm (Eds.), *The semantic web—ISWC 2010* (pp. 96–111). Berlin Heidelberg: Springer.

Calbimonte, J.-P., Jeung, H. Y., Corcho, O., & Aberer, K. (2012). Enabling query technologies for the semantic sensor web. *International Journal on Semantic Web and Information Systems, 8*(1), 43–63.

Chen, G., Jiang, T., Wang, M., Tang, X., & Ji, W. (2020). Modeling and reasoning of IoT architecture in semantic ontology dimension. *Computer Communications*, *153*, 580–594. https://doi.org/10.1016/j.comcom.2020.02.6.

Chu, J., Fu, H., Gao, F., & Zhao, D. (2017). Towards complex event processing in linked data stream. In *2017 12th IEEE conference on industrial electronics and applications (ICIEA)* (pp. 1016–1021).

Cugola, G., & Margara, A. (2012). Processing flows of information: From data stream to complex event processing. *ACM Computing Surveys (CSUR)*, *44*(3), 15.

D'Aniello, G., Gaeta, M., & Orciuoli, F. (2018). An approach based on semantic stream reasoning to support decision processes in smart cities. *Telematics and Informatics*, *35*(1), 68–81. https://doi.org/10.1016/j.tele.2017.09.019.

Dao-Tran, M., & Le Phuoc, D. (2015). Towards enriching cqels with complex event processing and path navigation. In *HiDeSt@KI* (pp. 2–14).

Drools fusion documentation. https://docs.jboss.org/drools/release/6.2.0.CR3/drools-docs/html/DroolsComplexEventProcessingChapter.html.

Fobel, A., & Subramanian, N. (2016). Comparison of the performance of drools and jena rule-based systems for event processing on the semantic web. In *2016 IEEE 14th international conference on software engineering research, management and applications (SERA)* (pp. 24–30).

Grosan, C., & Abraham, A. (2011). *Rule-based expert systems* (pp. 149–185). Berlin, Heidelberg: Springer Berlin Heidelberg. https://doi.org/10.1007/978-3-642-21004-4_7.

Hassine, M. B. *Easy-rules github page*. https://github.com/j-easy/easy-rules.

Helmer, S., Poulovassilis, A., & Xhafa, F. (2011). *Reasoning in event-based distributed systems. Vol. 347* (pp. 51–64). Springer.

Hoeksema, J., & Kotoulas, S. (2011). High-performance distributed stream reasoning using S4. In *Ordering workshop at ISWC*.

Hu, S., Huang, M., Feng, W., & Zhang, Y. (2017). A smart health service model for elders based on eca-s rules. In *2017 IEEE 15th international conference on software engineering research, management and applications (SERA)* (pp. 93–97).

Karaagac, A., Dalipi, E., Crombez, P., De Poorter, E., & Hoebeke, J. (2019). Light-weight streaming protocol for the internet of multimedia things: Voice streaming over NB-IoT. *Pervasive and Mobile Computing*, *59*. https://doi.org/10.1016/j.pmcj.2019.101044.

Kayan, H., Majib, Y., Alsafery, W., Barhamgi, M., & Perera, C. (2021). AnoML-IoT: An end to end re-configurable multi-protocol anomaly detection pipeline for Internet of Things. *Internet of Things*. https://doi.org/10.1016/j.iot.2021.1437.

Lephuoc, D. *Official java implementation of the cqels execution framework*. https://github.com/cqels/CQELS4J.

Le-Phuoc, D., Dao-Tran, M., Parreira, J. X., & Hauswirth, M. (2011). A native and adaptive approach for unified processing of linked streams and linked data. In *International Semantic Web Conference* (pp. 370–388). Springer.

Le-Phuoc, D., & Eiter, T. (2020). An adaptive semantic stream reasoning framework for deep neural networks. In *Proceedings of the CIKM 2020 workshops, October 19–20, Galway, Ireland*.

Le-Tuan, A., Hayes, C., Hauswirth, M., & Le-Phuoc, D. (2020). Pushing the scalability of RDF engines on IoT edge devices. *Sensors*, *20*(10), 2788. https://doi.org/10.3390/s20102788.

Lim, S., & Rahmani, R. (2020). Toward semantic IoT load inference attention management for facilitating healthcare and public health collaboration: A survey. *Procedia Computer Science*, *177*, 371–378. https://doi.org/10.1016/j.procs.2020.10.050.

Luo, X., Ye, F., Yin, L., Xun, H., & Li, Y. (2021). A scalable rule engine system for trigger-action application in large-scale IoT environment. *Computer Communications*, *177*, 220–229. https://doi.org/10.1016/j.comcom.2021.06.016.

Margara, A., Urbani, J., van Harmelen, F., & Bal, H. (2014). Streaming the web: Reasoning over dynamic data. *Web Semantics: Science, Services and Agents on the World Wide Web, 25*, 24–44. http://www.sciencedirect.com/science/article/pii/S157082681467.

Mauri, A., Calbimonte, J.-P., Dell'Aglio, D., Balduini, M., Brambilla, M., Della Valle, E., & Aberer, K. (2016). Triplewave: Spreading rdf streams on the web. In P. Groth, E. Simperl, A. Gray, M. Sabou, M. Krötzsch, F. Lecue, F. Flock, & Y. Gil (Eds.), *The semantic web—ISWC 2016* (pp. 140–149). Cham: Springer International Publishing.

McCarthy, D., & Dayal, U. (1989). The architecture of an active database management system. *SIGMOD Record, 18*(2), 215–224. http://doi.acm.org/10.1145/66926.66946.

Nguyen-Duc, M., Tuan Anh, L., Calbimonte, J.-P., Hauswirth, M., & Phuoc, D. (2020). *Autonomous RDF stream processing for IoT edge devices*. https://doi.org/10.1007/978-3-030-41407-8_20.

Official drools website. Red Hat Inc. https://www.drools.org/.

Parsia, B., Hitzler, P., Patel-Schneider, P., Krötzsch, M., & Rudolph, S. (2012). *OWL 2 web ontology language primer* (2nd ed.). W3C. Tech. Rep http://www.w3.org/TR/2012/REC-owl2-primer-20121211/.

Parsia, B., Motik, B., & Patel-Schneider, P. (2012). *OWL 2 web ontology language structural specification and functional-style syntax* (2nd ed.). W3C. W3C Recommendation http://www.w3.org/TR/2012/REC-owl2-syntax-20121211/.

Raimond, Y., & Schreiber, G. (2014). *RDF 1.1 primer, W3C, W3C Note*. http://www.w3.org/TR/2014/NOTE-rdf11-primer-20140624/.

Ren, X. (2018). *Distributed RDF stream processing and reasoning. Software Engineering*. Université Paris-Est. https://tel.archives-ouvertes.fr/tel-02083973.

Ren, X., Khrouf, H., Kazi-Aoul, Z., Chabchoub, Y., & Curé, O. (2016). *On measuring performances of C-SPARQL and CQELS*. SR+SWIT@ISWC (pp. 1–12).

Rhayem, A., Mhiri, M. B. A., & Gargouri, F. (2020). Semantic web technologies for the internet of things: Systematic literature review. *Internet of Things, 11*. https://doi.org/10.1016/j.iot.2020.1206.

Sundara, S., Das, S., & Cyganiak, R. (2012). *R2RML: RDB to RDF mapping language*. W3C. W3C Recommendation http://www.w3.org/TR/2012/REC-r2rml-20120927/.

Tanque, M. (2021). Chapter 2—Knowledge representation and reasoning in AI-based solutions and IoT applications. In G. Kaur, P. Tomar, & M. Tanque (Eds.), *Artificial intelligence to solve pervasive Internet of Things issues* (pp. 13–49). Academic Press. https://doi.org/10.1016/B978-0-12-818576-6.2-2.

Tommasini, R., Bonte, P., Ongenae, F., & Della, V. E. (2021). RSP4J: An API for RDF stream processing. In R. Verborgh, et al. (Eds.), *ESWC 2021. Lecture notes in computer science: Vol. 12731. The Semantic Web*. Cham: Springer. https://doi.org/10.1007/978-3-030-77385-4_34.

Valero, C. I., Medrano Gil, A. M., Gonzalez-Usach, R., Julian, M., Fico, G., Arredondo, M. T., Stavropoulos, T. G., Strantsalis, D., Voulgaridis, A., Roca, F., Jara, A. J., Serrano, M., Zappa, A., Khan, Y., Guillen, S., Sala, P., Belsa, A., Votis, K., & Palau, C. E. (2021). AIoTES: Setting the principles for semantic interoperable and modern IoT-enabled reference architecture for Active and Healthy Ageing ecosystems. *Computer Communications, 177*, 96–111. https://doi.org/10.1016/j.comcom.2021.06.010.

W3 n.d. RDF stream processing community group. https://www.w3.org/community/rsp/wiki/RDF_Stream_Processors_Implementation.

Xhafa, F., Kilic, B., & Krause, P. (2020). Evaluation of IoT stream processing at edge computing layer for semantic data enrichment. *Future Generation Computer Systems, 105*, 730–736. https://doi.org/10.1016/j.future.2019.12.031.

CHAPTER THREE

State-of-the-art research and development on Smart City

The concept of "Smart City" has been a proliferating topic in the past two decades. Among the widespread discussion on this concept, two main features could be observed. The first one aims to leverage information and communications technology (ICT) infrastructures and innovative solutions to tackle citywide issues, including economic development, mobility facilitation, environmental protection, and city governance. The second aims to improve the quality of life of the citizens and enhance the city's sustainability and competitiveness. Despite the extensive discussion and research on Smart City, there is no universally adopted definition for Smart City.

3.1 Definitions of Smart City

Before discussing the definitions of Smart City, it should be noticed that there are multiple concepts related to Smart City, such as Smart City, knowledge city, sustainable city, green city, etc. (Cocchia, 2014). Cocchia also pointed out the main difficulties in achieving a well-agreed definition of Smart City, which are

1. Various meanings are attributing to the word "smart," and many typologies of the city, such as Intelligent City, Wired City, Digital City and so on, could refer to Smart City.
2. Smart City is a fuzzy concept, and many cities define themselves as Smart City by some of their characteristics, but without a standard meaning.

As shown in Table 3.1, multiple city-related concepts are overlapping with Smart City (Camero & Alba, 2019). In addition, there also exist various definitions of Smart City due to the popularity and diversity of this concept. Albino et al. (2015) believed that Smart City was first used in the 1990s, and the focus was on the significance of new ICT about modern infrastructures within cities. However, some years later, the idea of a Smart City has been criticized as being too technically oriented. At the same time, Picon (2015)

Collective Intelligence for Smart Cities
https://doi.org/10.1016/B978-0-12-820139-8.00008-5

Table 3.1 City related concepts overlapping with Smart City.

City concept	References
Wired city	Dutton et al. (1987)
Eco city	Register (1987)
Sustainable city	Haughton and Hunter (2004)
Digital city	Ishida (2002)
Intelligent city	Komninos (2006)
Knowledge city	Edvinsson (2006)
Green city	OECD (2013)

stated that Smart City was coined around 2005 and referred to new ICTs and information processing capabilities to address urban design problems. Under the definition by Picon, Smart City is a self-aware and almost conscious entity with "internalized" mechanisms for learning, understanding, and reasoning through the ubiquitous deployment of sensors, monitors and data inputs, and the dynamic, real-time, and intelligent processing capabilities of information to identify solutions.

Nam and Pardo (2011) examined the meanings of the term "smart" in the Smart City context. They investigated the smartness of a city in marketing, urban planning, and technology aspects. For instance, smartness in the technology context implies the automatic computing principle like self-configuration, self-healing, self-protection, and self-optimization. Meanwhile, smart buildings like airports, hospitals, or university campuses are equipped with a multitude of mobile terminals and embedded devices, and connected sensors and actuators. The authors pointed out that "Smart City integrates technologies, systems, infrastructures, services, and capabilities into an organic network."

On the other hand, Camero and Alba (2019) emphasized that Smart City should not just talk about technology but also consider its citizens, "it is important to stress that cities are human settlements and they are supposed to serve us." Caragliu et al. (2011) believed the quality of life is an essential element to determine the smartness of a city. They mentioned that "a city to be smart when investments in human and social capital and traditional (transport) and modern (ICT) communication infrastructure. The investments fuel sustainable economic growth and high quality of life, with a wise management of natural resources, through participatory governance." Similar concept has also been raised by Pašalić et al. (2020) as they agreed that Smart City needs to "provide advanced and innovative services to its citizens, impacting the quality of life and sustainable management of natural resources." Batty et al. (2012) further stressed that a city could only be smart if it can improve the efficiency, equity, sustainability, and quality of life in the city.

Some definitions of Smart City are listed as follows. The list is the combination of definitions summarized by some studies (Albino et al., 2015; Camero & Alba, 2019; Yigitcanlar et al., 2019), and it should be pointed out that the first four of the definitions are reported by Camero and Alba (2019) as the most cited definition for Smart City.

- A city that monitors and integrates conditions of all of its critical infrastructures, including roads, bridges, tunnels, rails/subways, airports, seaports, communications, water, power, even major buildings, can better optimize its resources, plan its preventive maintenance activities, and monitor security aspects while maximizing services to its citizens... systems and structures will monitor their conditions and carry out self-repair, as "needed" (Hall et al., 2000).
- A Smart City is a city well performing in a forward-looking way in these six characteristics, built on the "smart" combination of endowments and activities of self-decisive, independent, and aware citizens (Giffinger et al., 2007).
- It is a city connecting the physical infrastructure, the IT infrastructure, the social infrastructure, and the business infrastructure to leverage the collective intelligence of the city (Harrison et al., 2010).
- A city to be smart when investments in human and social capital and traditional (transport) and modern (Information and Communications Technology, ICT) communication infrastructure fuel sustainable economic growth and high quality of life, with a wise management of natural resources, through participatory governance (Caragliu et al., 2011).
- Smart community—a community that makes a conscious decision to aggressively deploy technology as a catalyst to solving its social and business needs—will undoubtedly focus on building its high-speed broadband infrastructures. Still, the real opportunity is in rebuilding and renewing a sense of place and, in the process, a sense of civic pride (Eger, 2009).
- Using Smart Computing technologies to make the critical infrastructure components and services of a city—including city administration, education, healthcare, public safety, real estate, transportation, and utilities—are more intelligent, interconnected, and efficient (Washburn et al., 2009).
- Cities as territories are with high capacity for learning and innovation, which is built in the creativity of their population, their institutions of

knowledge creation, and their digital infrastructure for communication and knowledge management (Komninos, 2011).

- A Smart City is understood as a certain intellectual ability that addresses several innovative socio-technical and socio-economic aspects of growth. These aspects lead to Smart City conceptions as "green," referring to urban infrastructure for environmental protection and reduction of CO_2 emission, "interconnected" related to the revolution of broadband economy, "intelligent" declaring the capacity to produce added value information from the processing of city's real-time data from sensors and activators. In contrast, the terms "innovating," "knowledge" cities interchangeably refer to the city's ability to raise innovation based on knowledgeable and creative human capital (Zygiaris, 2013).
- Smart cities initiatives try to improve urban performance by using data, information, and information technologies (IT) to provide more efficient services to citizens, to monitor and optimize existing infrastructure, to increase collaboration among different economic actors, and to encourage innovative business models in both the private and public sectors (Marsal-Llacuna et al., 2015).
- A Smart City is an urban locality functioning as a healthy system of systems with sustainable and balanced practices of economic, societal, environmental, and governance activities generating desired outcomes and futures for all humans and nonhumans (Yigitcanlar et al., 2019).

3.2 Characteristics of Smart City

As Albino et al. (2015) mentioned, the difficulty of reaching a general agreement on the definition of Smart City may be the concept emphasis on both hard domains and soft domains. Hard domains are referred to buildings, energy grids, mobility, logistics, etc., while soft domains are implied to education, social inclusion, governance, etc. Thus, the different definitions may focus on different features of the Smart City. This section will discuss what characteristics a Smart City should have.

Based on the previous research on Smart City, Nam and Pardo (2011) re-categorized the components of Smart City into three core categories, which are technology (infrastructures of hardware and software), people (creativity, diversity, and education), and institution (governance and policy). They

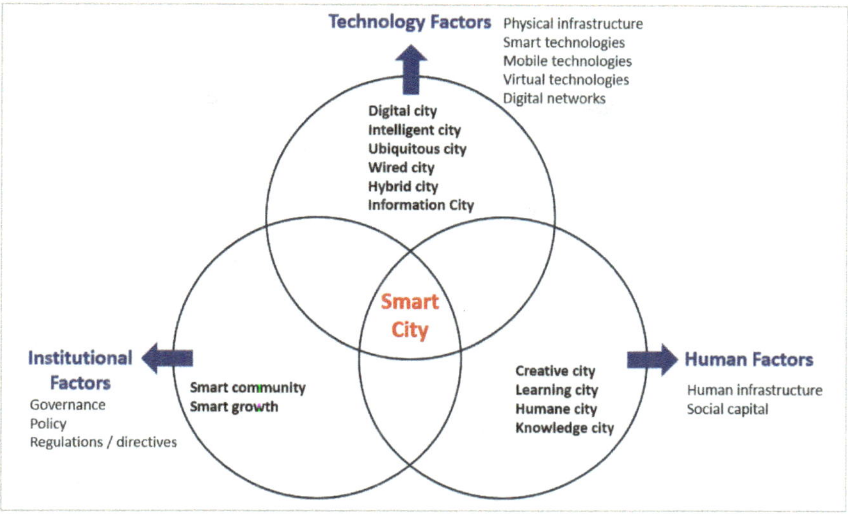

Fig. 3.1 Core components of Smart City.

pointed out that a city is smart when investments in human/social capital and IT infrastructure fuel sustainable growth and enhance the quality of life through participatory governance. Fig. 3.1 illustrates the components of Smart City proposed in this study. On the other hand, six "smart" characteristics, namely smart economy, smart people, smart governance, smart mobility, smart environment, and smart living, are identified for the Smart City by other studies (Giffinger & Gudrun, 2010; Lombardi et al., 2012). Giffer and Gudrun further broke down the six characteristics into 33 relevant factors to reflect the most important aspects for each characteristic. Table 3.2 lists the characteristics and their relevant factors identified by the authors.

The quality of life is the main component emphasized by many researchers (Giffinger & Gudrun, 2010; Lombardi et al., 2012; Shapiro, 2006). They stressed that Smart City should improve the living quality of its citizens. In particular, the infrastructures connection, economic development, and institutional governance in the Smart City should jointly enhance the sustainability of human citizens' city and life quality. Table 3.3 summarizes key components of Smart City in some previous research reported by Albino et al. (2015). Again, it could be noticed that people's life quality and social inclusion are treated as important factors, as shown in the list.

Table 3.2 Characteristics and factors of Smart City.

1. *Smart economy (competitiveness)* – Innovative spirit – Entrepreneurship – Economic image and trademarks – Productivity – The flexibility of the labor market – International embeddedness – Ability to transform	2. *Smart people (social and human capital)* – Level of qualification – Affinity to lifelong learning – A social and ethnic plurality – Flexibility – Creativity – Cosmopolitanism/Open-mindedness – Participation in public life
3. *Smart governance (participation)* – Participation in decision-making – Public and social services – Transparent governance – Political strategies and perspectives	4. *Smart mobility (transport and ICT)* – Local accessibility – (Inter-)national accessibility – Availability of ICT-infrastructure – Sustainable, innovative, and safe transport systems
5. *Smart environment (natural resources)* – Lack of pollution of natural conditions – Pollution – Environmental protection – Sustainable resource management	6. *Smart living (quality of life)* – Cultural facilities – Health conditions – Individual safety – Housing quality – Education facilities – Touristic

Table 3.3 Key components of Smart City in some previous research.

Source	Key components of Smart City
Mahizhnan (1999)	– IT education – IT infrastructure – IT economy – Quality of life
Eger (2009)	– Technology – Economic development – Job growth – Increased quality of life
Thuzar (2011)	– Quality of life – Sustainable economic development – Management of natural resources through participatory policies – Convergence of economic, social, and environmental goals

Table 3.3 Key components of Smart City in some previous research—cont'd

Source	Key components of Smart City
Chourabi et al. (2012)	– Management and organizations – Technology – Governance – Policy context – People and communities – Economy – Built infrastructure – Natural environment
Barrionuevo et al. (2012)	– Economic (GDP, international transactions, foreign investment) – Human (talent, innovation, creativity, education) – Social (traditions, habits, religions, families) – Environmental (energy policies, waste and water management, landscape) – Institutional (civic engagement, administrative authority, elections)

3.3 Measurement for Smart City

The development of Smart City has endeavored for decades in many places. Thus, it is important to evaluate the performance of Smart City. Zygiaris (2013) developed a framework, named Smart City Reference Model, to assess the smartness of a city. There are seven layers in the proposed model. The city layer is layer 0, which emphasizes the development of a Smart City must be grounded in the context of "city" (instead of "smart"). The green city layer is layer 1, which emphasizes the importance of the environment and sustainability. Layer 2 is the interconnection layer which emphasizes the citywide diffusion of green economies and interconnection of devices, buildings, people, etc. The instrumentation layer is layer 3, which emphasizes the real-time response of the system integrated by transmitters, smart meters, sensors, etc. The open integration layer is layer 4 related to the intercommunication and data sharing among Smart City applications. The application layer is layer 5, which emphasizes the real-time city operation reflected and integrated through systems. Finally, the innovation layer is layer 6, which focuses on the creation of a fertile innovation environment for new business opportunities. Utilizing this model, the author evaluated the performance of a Smart City as the case study (e.g., Barcelona).

Indicator assessment is another approach to measure the performance of Smart City. Lombardi et al. (2012) identified 60 indicators under five clusters to evaluate the performance of Smart City. The clusters are smart governance (related to participation), smart human capital (related to people), smart environment (related to natural resources), smart living (related to the quality of life), and smart economy (related to competitiveness). Carli et al. (2013) also adopted indicators to measure the smartness of a city in Italy (Bari), and they further proposed to classify the performance indicators of Smart City into two dimensions: objective and subjective.

Comparison and ranking are other commonly adopted methods to measure the performance of Smart City. By adopting 74 indicators and allocating them into six characteristics, Giffinger et al. (2007) developed an evaluation metric and managed to rank 70 medium-size cities in Europe for the city's smartness. Under their model, each city was marked in 74 indicators belonging to the 31 predefined factors, and every factor would be allocated to the six characteristics defined by their study. Thus, the performance information for each characteristic per city would be summarized. The summarized information is used to produce a list of rankings among these 70 cities. Table 3.4 shows the results of the 70 selected cities for their performance as a Smart City. City ranking has been conducted in many studies and organizations to evaluate the smartness for cities, for instance, IESE Cities in Motion Index, IMD Smart City Index, and Smart Cities Competitive Assessment. However, it should be pointed out that the ranking results may differ due to the difference in the evaluation standards and methods.

Table 3.4 Final results of the selected cities for their performance as Smart City.

CC	City	Smart economy	Smart people	Smart governance	Smart mobility	Smart environment	Smart living	Rank
LU	Luxembourg	1	2	13	6	25	6	1
DK	Aarhus	4	1	6	9	20	12	2
FI	Turku	16	8	2	21	11	9	3
DK	Aalborg	17	4	4	11	26	11	4
DK	Odense	15	3	5	5	50	17	5
FI	Tampere	29	7	1	27	12	8	6
FI	Oulu	25	6	3	28	14	19	7
NL	Eindhoven	6	13	18	2	39	18	8
AT	Linz	5	25	11	14	28	7	9
AT	Salzburg	27	30	8	15	29	1	10
FR	Montpellier	30	23	33	24	1	16	11

Table 3.4 Final results of the selected cities for their performance as Smart City—cont'd

CC	City	Smart economy	Smart people	Smart governance	Smart mobility	Smart environment	Smart living	Rank
AT	Innsbruck	28	35	9	8	40	3	12
AT	Graz	18	32	12	17	31	5	13
NL	Nijmegen	24	14	14	3	51	24	14
NL	Groningen	14	9	15	20	37	13	15
BE	Gent	19	16	31	7	48	4	16
SI	Ljubljana	8	11	43	31	3	29	17
NL	Maastricht	26	18	17	1	43	14	18
SE	Joenkoeping	36	10	7	34	22	26	19
BE	Brugge	23	20	29	18	44	2	20
NL	Enschede	31	17	16	4	35	23	21
DE	Goettingen	11	34	20	12	15	31	22
SE	Umeaa	39	5	10	36	46	10	23
DE	Regensburg	9	40	27	19	38	22	24
FR	Dijon	38	29	22	26	9	25	25
FR	Nancy	41	31	23	25	10	20	26
DE	Trier	21	44	19	10	18	33	27
FR	Clermont-Ferrand	33	33	26	29	7	27	28
FR	Poitiers	48	37	28	33	8	15	29
SI	Maribor	49	21	37	40	2	32	30
IE	Cork	2	26	25	45	66	21	31
DE	Erfurt	32	47	21	13	21	45	32
DE	Magdeburg	47	50	35	22	17	39	33
DE	Kiel	45	45	48	16	23	38	34
HR	Zagreb	34	24	32	39	36	42	35
UK	Cardiff	13	39	44	38	60	30	36
UK	Leicester	3	42	49	32	64	40	37
UK	Portsmouth	7	38	47	35	63	43	38
UK	Aberdeen	10	28	42	42	67	35	39
EE	Tartu	40	15	30	47	49	60	40
ES	Pamplona	22	48	39	51	32	41	41
CZ	Plzen	43	49	61	30	54	28	42
ES	Valladolid	44	53	34	54	24	46	43
CZ	Usti Nad Labem	54	51	55	23	55	36	44
IT	Trento	20	57	24	65	30	48	45
PT	Coimbra	52	63	54	49	16	37	46
SK	Nitra	62	46	51	52	19	44	47
PL	Rzeszow	69	19	53	41	56	50	48
IT	Trieste	12	61	40	67	45	57	49
ES	Oviedo	37	55	38	44	68	34	50

Continued

Table 3.4 Final results of the selected cities for their performance as Smart City—cont'd

CC	City	Smart economy	Smart people	Smart governance	Smart mobility	Smart environment	Smart living	Rank
IT	Ancona	35	59	36	68	34	49	51
IT	Perugia	42	54	41	66	42	51	52
PL	Bialystok	67	22	59	56	47	55	53
SK	Kosice	66	43	50	48	53	52	54
RO	Timisoara	50	64	64	62	4	59	55
SK	Banska Bystrica	70	41	52	53	58	47	56
PL	Bydgoszcz	68	27	57	46	52	61	57
GR	Patrai	59	58	46	60	5	67	58
LT	Kaunas	55	36	66	55	27	65	59
GR	Larisa	61	60	45	63	6	66	60
HU	Gyor	46	68	62	37	41	63	61
PL	Szczecin	65	52	58	43	59	56	62
RO	Sibiu	57	65	60	64	13	62	63
PL	Kielce	63	56	56	57	62	54	64
HU	Pecs	56	62	65	58	65	53	65
LV	Liepaja	60	12	63	61	61	70	66
HU	Miskolc	58	67	67	50	70	58	67
RO	Craiova	64	66	68	70	33	64	68
BG	Pleven	51	70	69	69	57	69	69
BG	Ruse	53	69	70	59	69	68	70

3.4 Smart City implementation: From the view of IoT and Big Data Analytics

Building a Smart City is a comprehensive project as it may involve every aspect of the planning of a city and the life of its residents. Based on the framework developed by Rathore, Paul, et al. (2018), IoT and Big Data Analytics, as a fundamental component for a Smart City, would be discussed in detail in this section.

IoT technology and infrastructure are considered crucial for the city to move toward a Smart City as the technology is transforming the traditional elements of living in a city. For example, IoT might impact the safety, mobility, and accessibility of a citizen. By being integrated with other facilities in a city and connected to a cloud-based control system, IoT technology could provide citizens with unprecedented services and experiences, such as smart parking, smart firefighting, and smart building. The IoT infrastructure could also enable Smart City to tackle the issues in a broad area, for instance, the health system (Rathore, Son, et al., 2018).

The IoT-based Smart City developed by Rathore, Paul, et al. (2018) is illustrated in Fig. 3.2. On the one hand, this IoT-based digital system enables the city to obtain real-time and accurate city information in the right situation to perform decision-making. On the other hand, it provides convenience and safety to individuals in the city. As depicted by the authors, "the real-time city data is harvested from smart homes, smart parking, water and weather system, environment population system, security, and surveillance system, vehicular traffic, and others, and all of this real-time data is used by the authorities to make a city-related decision using intelligent building equipped with data analytics, tools, and algorithms."

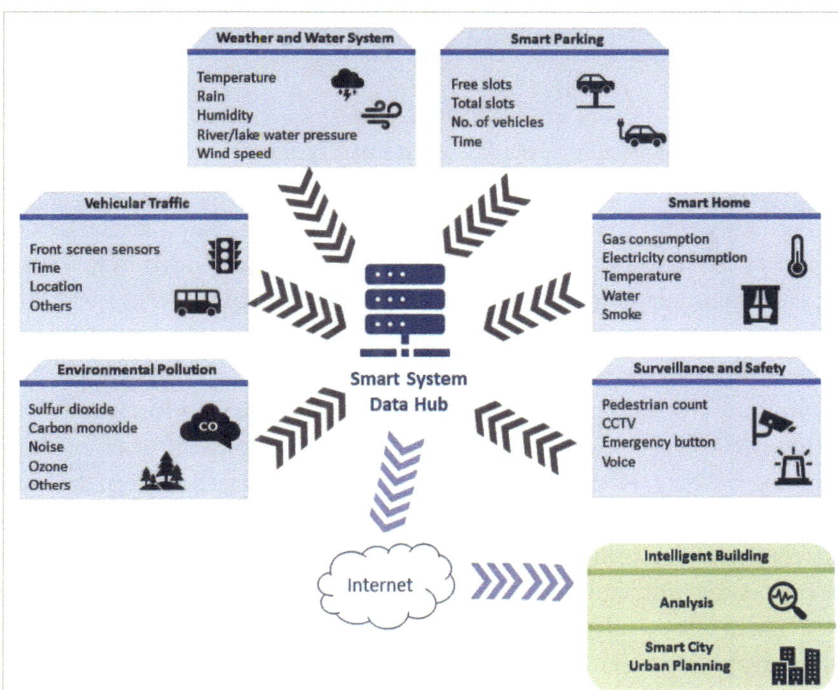

Fig. 3.2 IoT-based digital system of Smart City.

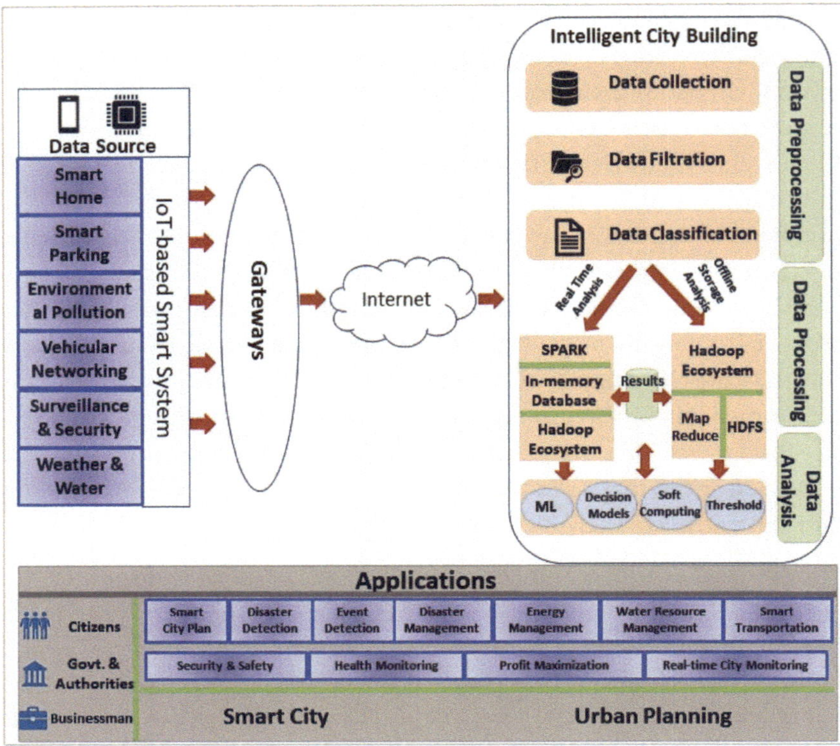

Fig. 3.3 System architecture of the IoT-based smart system.

Fig. 3.3 demonstrates how to implement the IoT-based digital system from data collection to decision making and services provided in the Smart City for the implementation aspect. Under this structure, every smart subsystem, such as smart home system, weather system, and smart parking system, will continuously produce, collect, and transfer real-time data continuously to the central smart system data hub, named Intelligent City Building (ICB). The data transfer is through gateways and the internet. After receiving the data, ICB will perform a set of activities starting from data collection and ending at decision-making while performing filtration, data processing, and classification in between them. Then the data is sent to the processing unit, equipped with the Spark and Hadoop ecosystem, to perform parameter calculation and other processing depending upon the algorithm given. In general, high-speed and high-volume data, namely big data, were collected and transferred by the IoT-based Smart system (left side). The big data analytics assembly (right side) would undertake processing and analysis to support the applications (bottom side) in the Smart City.

Challenges also exist during the implementation of smart cities. The scalability issue is the first challenge needed to be carefully addressed. The scalability of various techniques and applications should be well designed and planned due to the large scale of smart cities in various aspects, such as many populations, vehicles, buildings, services, etc. Following the scalability issue, another problem comes to the issue of reliability. The coexistence of multiple technologies could post great difficulties when various protocols communicate with each other. Therefore, some applications in the Smart City may fail and result in serious consequences to citizens. For example, the failures in vehicular applications, emergency response systems, or disaster management, etc., would cause severe consequences to people and society. The privacy issue is the next challenge for smart cities. As smart applications can track people and access personal data, privacy protection should be one of the top priorities during the development and implementation of Smart City applications. For instance, ensuring cybersecurity and preventing illegal access to personal data should be carefully handled. The final challenge is about public acceptance for Smart City applications. If the Smart City applications are not well recognized and accepted by the public, it would hinder the adoption of Smart City as many applications require several legal and administrative approvals. In addition, more efforts may need to gain supports from some specific groups of people, such as the elderly, children, and impaired people. Besides the above systematic challenges, the challenges in specific technologies, such as electric vehicles, unmanned aerial vehicles, and smart buildings (Beneicke et al., 2019), are also needed to be closely analyzed and addressed.

3.5 Practices of smart cities

Practices have been made in many cities around the world to transform toward the Smart City. As mentioned in the previous sections, there are no commonly agreed definitions and standards on Smart City. Thus, different studies or organizations may define or rank a different list for Smart City. For example, Intelligent Community Forum (ICF) annually announces and awards its listed smart cities (named Smart21 Communities). The winning city earns high scores in five areas: broadband connectivity, knowledge workforce, digital inclusion, innovation, and marketing and advocacy (Nam & Pardo, 2011). Table 3.5 presents the list of Smart21 Communities from 2018 to 2020 (ICF, 2020).

Table 3.5 Smart21 communities awarded by ICF from 2018 to 2020.

Smart21 communities of 2020	Smart21 communities of 2019	Smart21 communities of 2018
Adelaide, South Australia, Australia	Abbotsford, British Columbia, Canada	Adelaide, South Australia, Australia
Binh Duong Smart City, Vietnam	Adelaide, South Australia, Australia	Chiayi City, Taiwan
Chiayi City, Taiwan	Binh Duong Smart City, Vietnam	Espoo, Finland
Curitiba, Paraná, Brazil	Chiayi City, Taiwan	Hamilton, Ontario, Canada
Fredericton, New Brunswick, Canada	Chicago, Illinois, United States	Hudson, Ohio, United States
Hamilton, Ontario, Canada	Curitiba, Paraná, Brazil	Ipswich, Queensland, Australia
Hudson, Ohio, United States	Greater Victoria, British Columbia, Canada	Issy-les-Moulineaux, France
Issy les Moulineaux, France	Hudson, Ohio, United States	Kelowna, British Columbia, Canada
Leeds, West Yorkshire, United Kingdom	Issy-les-Moulineaux, France	Kinmen County, Taiwan
Markham, Ontario, Canada	Keelung City, Taiwan	Niagara Falls, Ontario, Canada
Matsu, Taiwan	Moscow, Russia	Olds, Alberta, Canada
Newmarket, Ontario, Canada	Nairobi County, Kenya	Parkland County, Alberta, Canada
Philadelphia, Pennsylvania, United States	Prospect, South Australia, Australia	Pickering, Ontario, Canada
Prospect, South Australia, Australia	Rochester, New York, United States	Sarnia-Lambton County, Ontario, Canada
Rochester, New York, United States	Sarnia-Lambton County, Ontario, Canada	St. Albert, Alberta, Canada
Sunshine Coast, Queensland, Australia	Sunshine Coast, Queensland, Australia	Sunshine Coast, Queensland, Australia
Tallinn, Estonia	Surat, Gujarat, India	Tainan City, Taiwan
Wellington, New Zealand	Tainan City, Taiwan	Taoyuan, Taiwan
Westerville, Ohio, United States	Taoyuan, Taiwan	Western Valley, Nova Scotia, Canada
Whanganui, New Zealand	Westerville, Ohio, United States	Winnipeg, Manitoba, Canada
Winnipeg, Manitoba, Canada	Winnipeg, Manitoba, Canada	York, Ontario, Canada

Yigitcanlar et al. (2019) examined five of the role model smart cities, which are Songdo (Korea), Masdar (UAE), Amsterdam (the Netherlands), San Francisco (United States), and Brisbane (Australia), to examine the real-world practice of Smart City. These five cities are selected from different regions globally and are referred to good practices in literature and have distinctive characteristics and visions in Smart City. The authors closely evaluated these five cities in city background, technology, community, policy, and global implications. Table 3.6 elaborates the detailed practices of these five selected smart cities.

Smart City is a popular topic during the world cities are transforming into sustainable cities. Despite some promising practices of smart cities, the notion of a Smart City still needs further conceptualized and implemented. Insights generated from the current practices could guide us in future

Table 3.6 Smart City practices of five selected cities.

Songdo	
Background	– An exemplar new city based on the Smart City concept – A large-scale greenfield-based Smart City project
Technology	– Filled with cutting-edge technologically enhanced infrastructure and services – A robust public transit system and an effective water recycling system – Aims to excel particularly in bio, nano, information, and ubiquitous technologies
Community	– The top-down development strategy without the involvement of all stakeholders – Can only serve to those who can afford and becoming a city for the affluent class only – Walkability and highly efficient clientele politics of the city
Policy	– Master plan is based on a combination of sustainable design principles – Cutting-edge urban technologies linked with sustainable urban design practices
Global implications	– Considered as a role model of the Southeast Asian Smart City – Created a new development path for and setting the benchmark high for smart urbanism
Masdar	
Background	– A planned Smart City project situated in a desert location near Abu Dhabi – Claimed as the largest planned development in the world that solely uses renewable energy

Continued

Table 3.6 Smart City practices of five selected cities—cont'd

Technology	– Autonomous and electrified public transport system, and walking and cycling network – The design of the walls of the buildings reduces demand for air conditioning by 55% – Movement sensors in all buildings cutting consumption of electricity and water by 51% and 55%
Community	– Only 20% of the accommodation areas are assigned to the low-income workers – The city is largely occupied by affluent population, pointing to the exclusiveness of the city
Policy	– Top-down planning and design approach – One of the first attempts in constructing carbon-neutral cities – Not economically feasible
Global implications	– One of the few projects with innovative and ambitious Smart City plans on a large scale – It brings out an important question: transforming cities step-by-step into smart ones or building new ones from scratch at scale
Amsterdam	
Background	– Two principles to initiate the project: (a) Enable stakeholders to apply innovative technologies; (b) Stimulate behavioral change with end-users – A quadruple-helix partnership model between public, private, academia, and community
Technology	– Engage local communities to determine a bottom-up solution for the city and residents – Technology or solution that was successfully tested and developed in the pilot project would be commercialized and widely applied in the city
Community	– The Smart City platform established partnerships with over 80 partners – The platform has evolved into a facilitator of the Smart City community in the city
Policy	– Established and maintained strong linkages with a number of European Smart City initiatives – Shared data with the wider community and provides critical info on the transport environment – One of the most walkable and cyclable cities in the world
Global implications	– Being considered as a role model European Smart City – Concerns on Data management and analysis should be addressed appropriately

Table 3.6 Smart City practices of five selected cities—cont'd

San Francisco	
Background	– The city is home to a large number of internet-based companies – Offers large number of free Wi-Fi hotspots in various public locations
Technology	– Implemented many incentive programs that involve smart technology applications, such as "SF Energy Map" and "Energy Use Challenge" – 41% renewable energy powered – Home of ride-hailing service companies and a trial city for shared autonomous vehicle projects
Community	– High concentration of talent base and strong entrepreneurial culture – Challenges to providing affordable housing to its residents
Policy	– An ambitious goal of achieving zero-waste – The open data legislation that passed in 2009
Global implications	– Considered as one of the greenest cities and the clean-technology capital of North America – Showcased a successful model of urban transformation – Came with social costs that require further attention and solid policies to tackle
Brisbane	
Background	– One of the early adopters of the Smart City concept – The knowledge corridor, a milestone project connecting all key innovative institutes of the city physically, is highly active, its global innovation districts are international recognition
Technology	– High technology smart traffic systems are utilized in road infrastructure including digital message signs, CCTV cameras, and Bluetooth sensor devices – Free Wi-Fi systems are installed in the major facilities, and followed by the erection of smart poles in the major public spaces for big data collection
Community	– Developing mechanisms for community involvement in major urban policy decisions – Be committed to the smart community development program
Policy	– Launched "CitySmart" to help make Brisbane Australia's most sustainable city – "CitySmart" is financially supported to deliver projects
Global implications	– As a role model, Oceanian Smart City – Lacked a clear overarching Smart City strategy to guide the transformation toward a smart and sustainable city

projects. One of the future Smart City planning focuses might be transforming from hardware-centric to more software-centric and people-centric.

References

Albino, V., Berardi, U., & Dangelico, R. M. (2015). Smart cities: Definitions, dimensions, performance, and initiatives. *Journal of Urban Technology*, *22*(1), 3–21.

Barrionuevo, J. M., Berrone, P., & Ricart, J. E. (2012). Smart cities, sustainable progress. *IESE Insight*, *14*(14), 50–57.

Batty, M., Axhausen, K. W., Giannotti, F., Pozdnoukhov, A., Bazzani, A., Wachowicz, M., Ouzounis, G., & Portugali, Y. (2012). Smart cities of the future. *The European Physical Journal Special Topics*, *214*(1), 481–518.

Beneicke, J., Juan, A. A., Xhafa, F., Lopez-Lopez, D., & Freixes, A. (2019). Empowering citizens' cognition and decision making in smart sustainable cities. *IEEE Consumer Electronics Magazine*, *9*(1), 102–108.

Camero, A., & Alba, E. (2019). Smart City and information technology: A review. *Cities*, *93*, 84–94.

Caragliu, A., Del Bo, C., & Nijkamp, P. (2011). Smart cities in Europe. *Journal of Urban Technology*, *18*(2), 65–82.

Carli, R., Dotoli, M., Pellegrino, R., & Ranieri, L. (2013). Measuring and managing the smartness of cities: A framework for classifying performance indicators. In *2013 IEEE international conference on systems, man, and cybernetics* (pp. 1288–1293). IEEE.

Chourabi, H., Nam, T., Walker, S., Gil-Garcia, J. R., Mellouli, S., Nahon, K., Pardo, T. A., & Scholl, H. J. (2012). Understanding smart cities: An integrative framework. In *2012 45th Hawaii international conference on system sciences* (pp. 2289–2297). IEEE.

Cocchia, A. (2014). Smart and digital city: A systematic literature review. In *Smart city* (pp. 13–43). Cham: Springer.

Dutton, W. H., Blumler, J. G., & Kraemer, K. L. (1987). *Wired cities: Shaping the future of communications*. GK Hall & Co.

Edvinsson, L. (2006). Aspects on the city as a knowledge tool. *Journal of Knowledge Management*, *10*(5), 6–13.

Eger, J. M. (2009). Smart growth, smart cities, and the crisis at the pump a worldwide phenomenon. *I-WAYS-The Journal of E-Government Policy and Regulation*, *32*(1), 47–53.

Giffinger, R., Fertner, C., Kramar, H., Kalasek, R., Pichler-Milanovic, N., & Meijers, E. (2007). *Smart cities—Ranking of European medium-sized cities*. Vienna University of Technology.

Giffinger, R., & Gudrun, H. (2010). Smart cities ranking: An effective instrument for the positioning of the cities? *ACE: Architecture, City and Environment*, *4*(12), 7–26.

Hall, R. E., Bowerman, B., Braverman, J., Taylor, J., Todosow, H., & Von Wimmersperg, U. (2000). The vision of a smart city. In *2nd International Life Extension Technology Workshop, Paris, France, September 28, 2000*.

Harrison, C., Eckman, B., Hamilton, R., Hartswick, P., Kalagnanam, J., Paraszczak, J., & Williams, P. (2010). Foundations for smarter cities. *IBM Journal of Research and Development*, *54*(4), 1–16.

Haughton, G., & Hunter, C. (2004). *Sustainable cities*. Routledge.

Intelligent Community Forum. (2020). *The Smart21 communities of the year*. Retrieved from https://www.intelligentcommunity.org/smart21. (Accessed May 2020).

Ishida, T. (2002). Digital city Kyoto. *Communications of the ACM*, *45*(7), 76–81.

Komninos, N. (2006). The architecture of intelligent cities. *Intelligent Environments*, *6*, 53–61.

Komninos, N. (2011). Intelligent cities: Variable geometries of spatial intelligence. *Intelligent Buildings International*, *3*(3), 172–188.
Lombardi, P., Giordano, S., Farouh, H., & Yousef, W. (2012). Modelling the smart city performance. *Innovation: The European Journal of Social Science Research*, *25*(2), 137–149.
Mahizhnan, A. (1999). Smart cities: The Singapore case. *Cities*, *16*(1), 13–18.
Marsal-Llacuna, M. L., Colomer-Llinàs, J., & Meléndez-Frigola, J. (2015). Lessons in urban monitoring taken from sustainable and livable cities to better address the smart cities initiative. *Technological Forecasting and Social Change*, *90*, 611–622.
Nam, T., & Pardo, T. A. (2011). Conceptualising smart city with dimensions of technology, people, and institutions. In *Proceedings of the 12th annual international digital government research conference: Digital government innovation in challenging times* (pp. 282–291).
OECD. (2013). Green growth in cities. In *OECD green growth studies* OECD Publishing. https://doi.org/10.1787/9789264195325-en.
Pašalić, I. N., Ćukušić, M., & Jadrić, M. (2020). Smart city research advances in Southeast Europe. *International Journal of Information Management*. https://doi.org/10.1016/j.ijinfomgt.2020.102127. Advance Online Publication.
Picon, A. (2015). *Smart cities: A spatialised intelligence*. John Wiley & Sons.
Rathore, M. M., Paul, A., Hong, W. H., Seo, H., Awan, I., & Saeed, S. (2018). Exploiting IoT and big data analytics: Defining smart digital city using real-time urban data. *Sustainable Cities and Society*, *40*, 600–610.
Rathore, M. M., Son, H., Ahmad, A., Paul, A., & Jeon, G. (2018). Real-time big data stream processing using GPU with spark over hadoop ecosystem. *International Journal of Parallel Programming*, *46*(3), 630–646.
Register, R. (1987). *Ecocity Berkeley: Building cities for a healthy future*. North Atlantic Books.
Shapiro, J. M. (2006). Smart cities: Quality of life, productivity, and the growth effects of human capital. *The Review of Economics and Statistics*, *88*(2), 324–335.
Thuzar, M. (2011). Urbanisation in southeast asia: Developing smart cities for the future? In M. J. Montesano, & P. O. Lee (Eds.), *Regional Outlook: Southeast Asia 2011–2012* (pp. 96–100). Singapore: ISEAS Publishing.
Washburn, D., Sindhu, U., Balaouras, S., Dines, R. A., Hayes, N., & Nelson, L. E. (2009). Helping CIOs understand "smart city" initiatives. *Growth*, *17*(2), 1–17.
Yigitcanlar, T., Han, H., Kamruzzaman, M., Ioppolo, G., & Sabatini-Marques, J. (2019). The making of smart cities: Are Songdo, Masdar, Amsterdam, San Francisco and Brisbane the best we could build? *Land Use Policy*, *88*, 104187.
Zygiaris, S. (2013). Smart city reference model: Assisting planners to conceptualise the building of smart city innovation ecosystems. *Journal of the Knowledge Economy*, *4*(2), 217–231.

CHAPTER FOUR

Internet of things (IoT), cloud computing, and big data collective intelligence for smart cities

4.1 Introduction of smart cities

In the last decades, the global population overgrew, so the expectation of the quality of life also has. It is forecasted that nearly 70% of the world population will live in urban areas by 2050. For the prediction of resource consumption, around 75% of the world's resources and energy will be consumed, resulting in 80% of greenhouse gases (Saraju et al., 2016). It reveals that in the long term, environmentally, our future would be worse. Despite the rapidly growing technology that can not completely deal with environmental problems, smart cities would be a way to control resource overuse. For instance, there would be a better use of information, digital, and telecommunication technologies to provide more flexible and efficient networks to people (Saraju et al., 2016). Smart cities provide more benefits in terms of environment and increase the quality of life, efficiency in projects regarding smart cities, and solving urban problems (Benevolo et al., 2016). Internet of things (IoT), cloud computing, and big data collection intelligence would be three key elements to complete and perfect smart cities. Hence, with those technologies, smart cities would bring positive effects to our society. This chapter would illustrate the details of IoT, cloud computing, and big data collection intelligence of smart cities.

4.2 Integration of IoT and cloud computing in smart cities

The IoT technology refers to network connectivity and computing capability. It allows objects to generate, exchange and consume accurate time information with minimal human intervention (Rose et al., 2015). Sensors would be a typical example of IoT. Also, everyday items that people

https://doi.org/10.1016/B978-0-12-820139-8.00016-4

would not consider as a computer would possibly become IoT. Generally, the architecture of traditional IoT is formed by three layers revealed in Fig. 4.1A (Muntjir et al., 2017):

(1) The application layer
(2) The network layer
(3) The perception layer

The application layer is at the top as data utilization in applications runs. For instance, applications include smart traffic, smart industry, environment protection, mining monitor, remote nursing, safety defense, and smart government. The network layer is in the middle, formed by output flow control and communication of devices. Mobile phone networks, fixed telephone networks, broadcasting networks, and closed IP data networks for each carrier are examples of this layer. The perception layer is to collect information about sensor objects. It would be considered the most critical driver of IoT.

Five-layer architecture in Fig. 4.1B is proposed owing to three-layer architecture becoming sufficient for the keep growing technologies and IoT development (Muntjir et al., 2017):

(1) Business layer
(2) Application layer
(3) Processing layer
(4) Transport layer and
(5) Perception layer

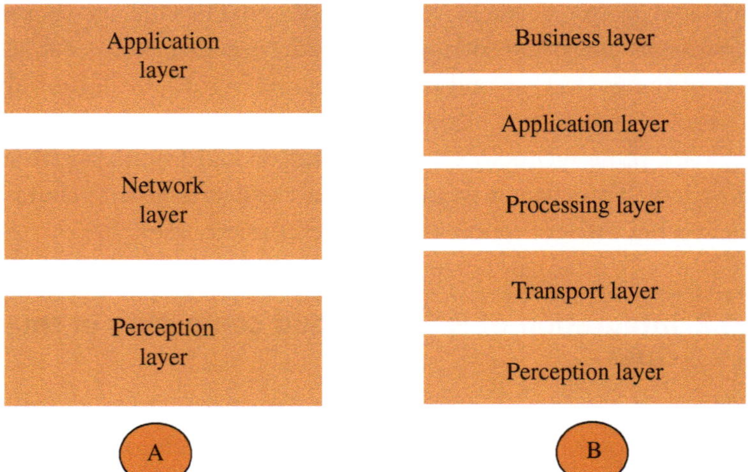

Fig. 4.1 Architecture of IoT (Muntjir et al., 2017).

The first layer at the top is responsible for all research related to IoT applications and users' privacy regarding change and management. Determination of types of applications used in the IoT would be the target of the Application layer, the second layer under business. The third layer, the processing layer, aims to handle processes such as storing and analyzing vast amounts of information. The following is the fourth one, the transport layer. Its responsibility is to transmit and receive the information from the perception layer to the processing layer by technologies like Wi-Fi or Bluetooth and vice versa. The last layer at the bottom is perception. In this layer, defining the physical meaning of each thing in the IoT system would be conducted. For example, gathering data about each object in the system and transformation of data to signals would be conducted.

Fig. 4.2 reveals some examples of IoT applications in smart cities.

In Smart homes, there are five typical applications (Malche & Maheshwary, 2017):

(1) Controlling the demand of users at homes such as opening doors or windows automatically.
(2) Detecting smoke particles in the air and giving alerts to prevent fire.
(3) Monitoring temperature via sensors to give real-time data or alert.
(4) Keep track of every activity around home and intelligence would make the best decision for users' security.
(5) Providing data of smart home such as illumination via social media like Facebook.

In Smart parking loT, there are five typical applications (Lookmuang et al., 2018):

(1) Detecting the number of vehicles within the range by camera and sensor.
(2) Knowing vacant parking slots before departure and finding parked locations after arrivals by mobile.
(3) Monitoring the car park environment to detect people or animals present in a range.
(4) Allowing users to check car status and current fee. It also allows users to pay the fee by mobile as a replacement for the physical ticket.
(5) Reducing congestion because people would not drive to an area of full parking.

There are four typical applications (El Zouka & Hosni, 2021; Niyato et al., 2009):

(1) Tracking of body parameters like temperature and internet history via the Internet.

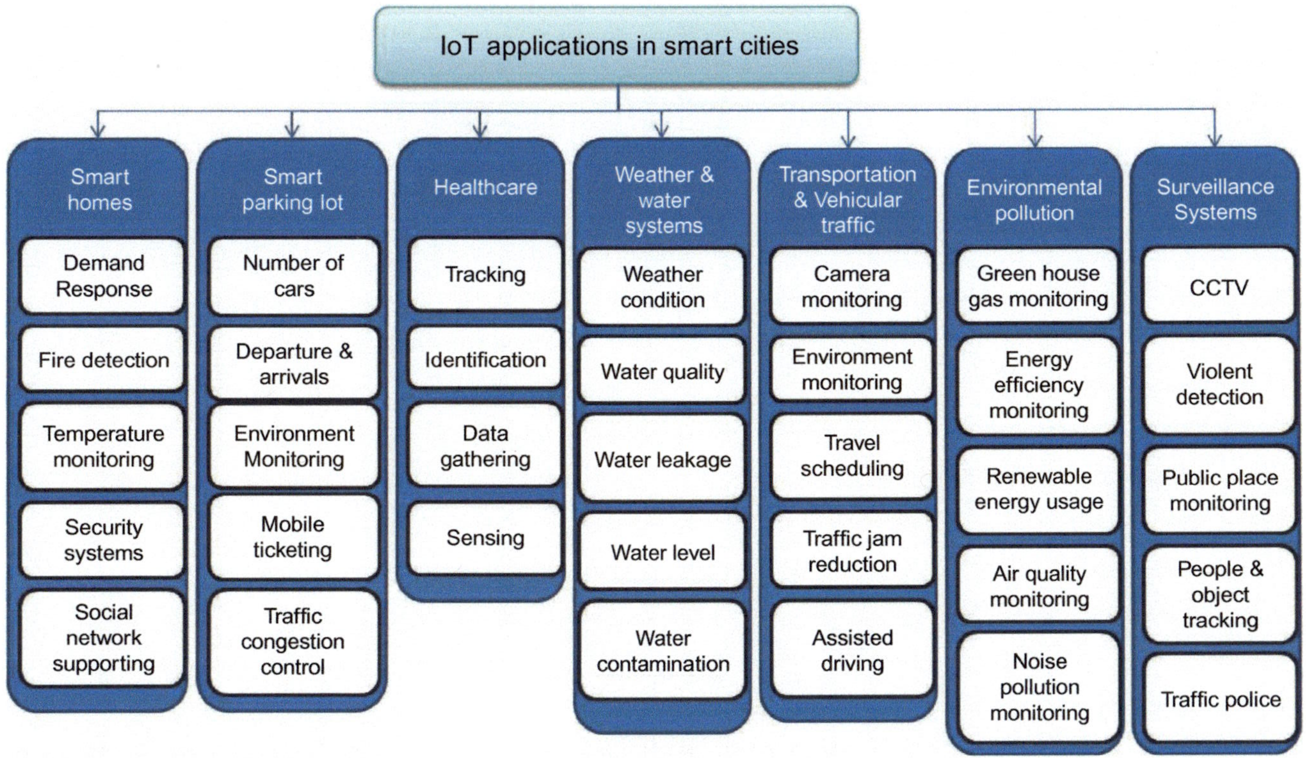

Fig. 4.2 The main applications of the IoT (Talari et al., 2017).

(2) Identification of patients' risk of mistakes or disease by drug or smoke histories.
(3) Allowing medical experts to collect, measure and analyze real-time data of patients. It also increases the ability of medical experts to handle urgent situations.
(4) Attaching sensors on the patients' body to sense and collect movement or body condition like heart rate or blood pressure in real-time.

There are five typical applications in Weather systems and Water systems (Badawi, 2019; Kodali & Mandal, 2016; Narendran et al., 2017; Pappu et al., 2017):

(1) Using sensors to get real-time weather conditions like wind speed, wind direction, and temperature is popular.
(2) Machines like Water tanks with sensors and processors would capture and analyze water quality in a residential area.
(3) Detecting water leakage in the underground pipeline and getting a warning message.
(4) Estimating groundwater levels and giving out alerts to the local community.
(5) Contamination level in water would be determined by laser beam on the algae.

In Transportation & Vehicular Traffic there are five typical applications (Hossain & George, 2018; Saifuzzaman et al., 2017):

(1) Installing cameras on roads allows controlling and monitoring via the Internet.
(2) Monitoring things in traffic environment like livings or lights inroads.
(3) Traffic jam status of roads would be shown so users can easily change the route for better travel scheduling.
(4) Installing barriers inroads that appear automatically in red signals to prevent traffic accidents. Lowering accidents may lower-traffic jam.
(5) Waking the driver up when the driver is drowsy by detecting the driver's eye.

In environmental pollution, there are five typical applications (Chen et al., 2018; Choi et al., 2018; Dhingra et al., 2019; Kim et al., 2020; Segura Garcia et al., 2016):

(1) Using sensors to collect data and then calculate greenhouse gas emissions in real-time anytime, anywhere. Users would monitor greenhouse gas emissions more effectively.

(2) Monitoring systems would provide real-time energy data. Users would have more information on energy consumption to conducting better policies for saving energy.
(3) Collecting data about energy status from solar and wind power generation facilities. Users would know how much renewable power is used and conduct better arrangements on renewable energy usage.
(4) Collecting data about gas in the air and sending those data to cloud for the machine to determine air pollution level and conduct prediction for air pollution.
(5) Noise data collections via IoT allowing users to calculate noise levels and conduct noise measurement.

In surveillance systems, there are five typical applications (Calavia et al., 2012; Damen & Hogg, 2012; Kim & Park, 2019; Wang & Zhang, 2016):

(1) Conducting face recognition and monitoring a particular area by CCTV to prevent crimes and know the criminals if there is a crime that cannot prevent.
(2) Monitoring people's actions to detect violent acts or even recognize the victim and crime.
(3) Monitoring the massive number of people in a public area by human temperature.
(4) Considering irregularities in people's silhouette by video sequence frameworks to find out possible banned carry objects.
(5) Aware traffic police if some cars are in a wrong direction by using running motion algorithms.

The rapid growth of technologies such as Sensor Technology, Radio Frequency Technology and collecting Information Technology, etc., boost the development of IoT (Madakam, Ramaswamy, & Tripathi, 2015). Radio Frequency Identification (RFID), Internet Protocol (IP), Electronic Product Code (EPC), Barcode, Wireless Fidelity (Wi-Fi), Bluetooth, or Artificial Intelligence (AI) would be the common examples of the above growing technologies. With that technology, people would automatically find out every kind of problem everywhere by machines, such as traffic, human health, crimes, or natural resources. Thus, with the help of technology, people would control and figure out trouble faster so that they would be able to enhance efficiency in decision making for dealing with the above issue quicker. Nonetheless, applying IoT in smart cities also exists challenges such as handling heterogeneous and massive data generated by smart cities with an IoT environment. Cloud computing, an internet-based computing technology, would be useful to cope with the difficulties of handling heterogeneous and massive data. Generally, cloud computing refers to internet-based computing that provides shared computer processing resources to digital entities such as servers, printers, storage, services, and applications. It provides on-demand network access and hardware and software resources by different on-demand models depending on different situations (Kaur & Maheshwari, 2016). Infrastructure-as-a-Service (Iaas), Platform-as-a-Service (PaaS), Software-as-a-Service (SaaS), Network-as-a-Service (NaaS), Storage-as-a-Service (STaaS), and Sensor-as-a-Service (SSaaS) would be the common examples of cloud models. Fig. 4.3 shows the responsibilities of the three most common service models, SaaS, PaaS, and IaaS.

The responsibilities of Vendors in different cloud models are different (Strømmen-Bakhtiar & Razavi, 2011). Vendors in SaaS have the largest number of responsibilities, such as software, platform, and infrastructure. Vendors in PaaS have platform and infrastructure as responsibilities. Vendors in IaaS only control the infrastructure. It would be concluded that the flexibility of IaaS is the highest for users, followed by PaaS, and the lowest is SaaS. The more responsibilities managed by vendors would reveal the less effort that users pay.

Different cloud models provide different services (Puthal et al., 2015). The SaaS model uses applications on cloud infrastructure via the web browser to access software. The PaaS model provides services of applications created by users who use a set of programming languages and tools supported by PaaS providers. Services of the IaaS model can provision processing, storage, networks, and other fundamental computing resources.

Cloud computing allows a pay-per-use-basis which means users pay the fee only when they use the system. In smart cities, the heterogeneous and massive data produced would be processed efficiently at a lower cost. In addition, cloud computing is convenient to users because it allows users access anytime, anywhere with any various devices. Hence, the quality of IoT applications in smart cities would be enhanced, and the cost of smart cities would be lower with cloud computing.

4.3 Big data in smart city

Data sets in all large companies are enormous, and they also cannot store all their historical data for long periods. Besides, traditional technologies for data storage are costly, with limited storage capacity and inflexible management tools (Oussous et al., 2018). Thus, big data, a new term initiated from the demand of analyzing massive business data by large companies

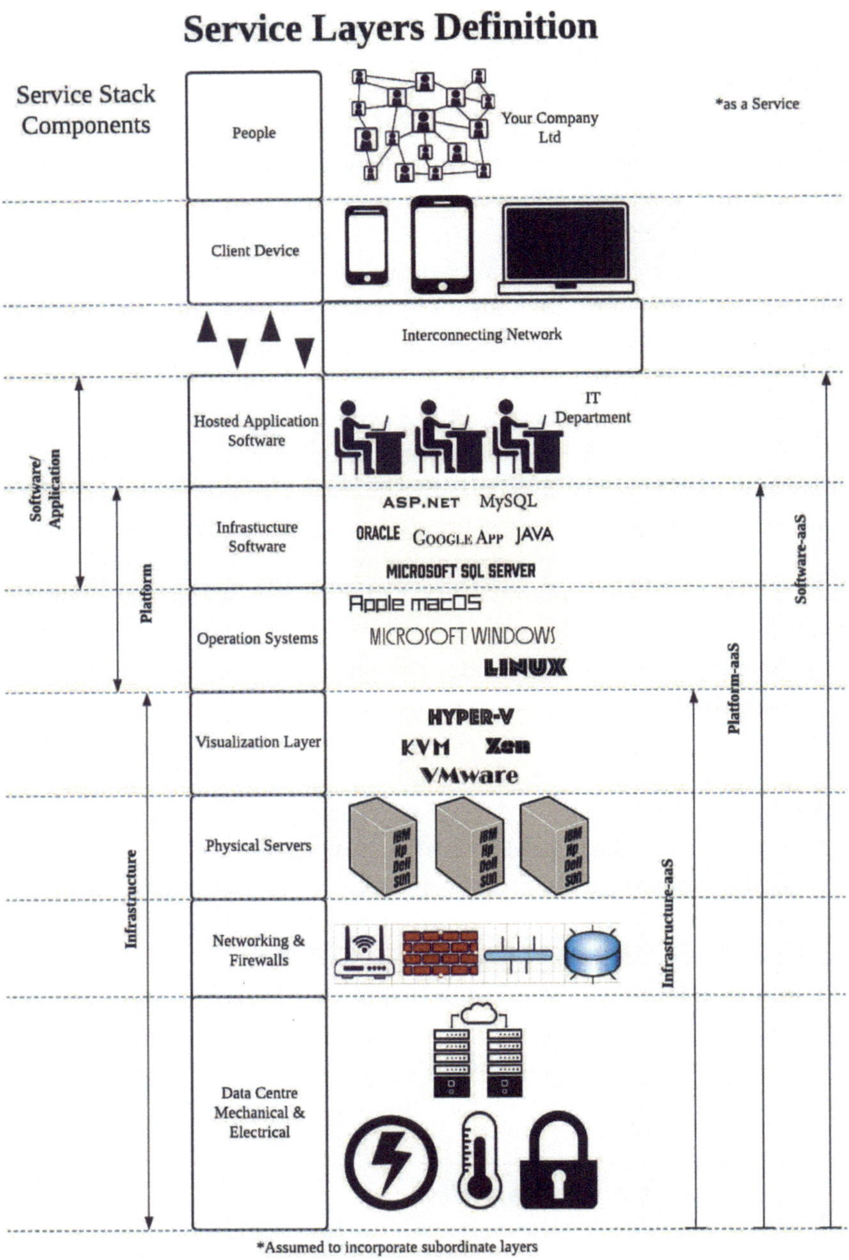

Fig. 4.3 IaaS vs PaaS vs SaaS (Strømmen-Bakhtiar & Razavi, 2011).

such as Yahoo, Google, and Facebook, have become popular (Yaqoob et al., 2016). Big data also include heterogeneous formats such as structured, unstructured, and semistructured data (Oussous et al., 2018). In detail, big data would be known as 3V, which shows its three main characteristics (Furht & Villanustre, 2016):

(1) Variety
(2) Velocity
(3) Volume

Variety means big data collected by different sources and various formats, such as videos, documents, comments, and logs. Structured, unstructured, semistructured data, public or private, local or distant, shared or confidential, complete or incomplete data would consist of big data together or separately. Velocity means big data generated and processed in a rapid way to fulfill users' needs. 2.5 PB of customers' transactions data would be generated every hour from Walmart, an international discount retail chain. Volume means big data contains a large volume of digital data from millions of devices and applications like ICTs, smartphones, products' codes, social networks, sensors, and logs. Fig. 4.4 reveals that two more Vs and other

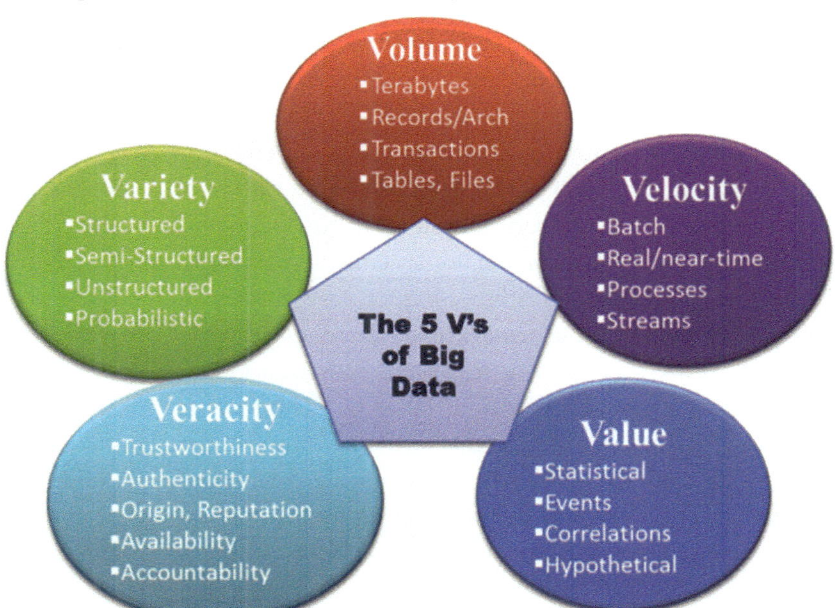

Fig. 4.4 The 5V's of big data (Dahdouh et al., 2020).

characteristics would be added to define big data better (Dahdouh, Dakkak, Oughdir, & Abdelali, 2020):

(4) Veracity

(5) Value

Veracity means big data should be captured correctly without miss or error. Data analysis would perform analysis via exact data, so the validity of big data would affect the decision making. Thus, techniques to eliminate useless data and ensure data having a high quality of validity are needed. Value means turning big raw data into meaningful information with value. Technologies and advanced methods such as DBMS or NoSQL would be helpful to perform this task for reducing effort and workload for business companies.

In the case of smart cities, data generated by IoT would be an example of big data. Every data captured within the smart cities regarding the environment, human health, traffic, and so on would be combined as big data in smart cities. Nevertheless, it would be not easy to collect, process, and analyze that massive data. In addition, after acquiring those big data, people would need to analyze those data with helpful data analysis techniques. So they can deal with their problem and conduct effective decision making in every area of smart cities. There are four types of data analytics for people to analyze big data (Rajaraman, 2016):

(1) Descriptive analytics

(2) Predictive analytics

(3) Exploratory or Discovery analytics and

(4) Prescriptive analytics

Descriptive analytics aim to show the past by bar charts, graphs, pie charts, maps, or scatter diagrams, which help visualize what the data implies easily. Predictive analytics infers from available data to tell the possible events in the future by time series analysis, including statistical methods, neural networks, and machine learning algorithms. Exploratory analytics finds unexpected relationships between big data. For instance, customers' behavior would be forecasted by their actions in tweets, blogs, Facebook data, or emails. If one cancels a news subscription, other news companies will try to offer something for that one. Prescriptive analytics optimizes solutions to existing problems by analyzing opportunities from big data. So users would know what to do for different problems or in different situations.

Furthermore, two bits of intelligence would assist users in conducting the above data analytics (Iqbal et al., 2020):

(1) Machine learning

(2) Computational Intelligence

Modeling patterns and correlations in data are the major offering of machine learning. Data analysis would discover relationships and make predictions based on unseen events via machine learning. Supervised learning such as learning from labeled data, unsupervised learning, and reinforcement learning are machine learning approaches. Computational Intelligence would cope with complex real-world data that traditional modelings cannot solve as a subclass of machine learning. It is because the complex real-world data are of high complexity, uncertainty, and stochastic nature of processes. The triad of core computational Intelligence like Fuzzy Logic (FL), Evolutionary Algorithms (EA), and Artificial Neural Networks (ANN) can deal with those complex real-world data.

Therefore, with big data analytics, people would analyze big data using different methods to get the different information they want. It is believed that people would achieve better in sustainability, resilience, and governance. Therefore, citizen's quality of life would be enhanced, and intelligent management of infrastructures and natural resources would be implemented (Al Nuaimi et al., 2015).

4.4 Application areas

As discussed before in Fig. 4.2, there are seven main applications of IoT related to smart cities. Some of those applications are widely discussed and implemented. In the following, those popular applications would be discussed in detail with the recommendation and benefits of integrating IoT, cloud computing, and extensive data analysis in smart cities.

Fig. 4.5 shows a proposed model of smart healthcare (Gupta et al., 2016). This model is a robust health monitoring system that monitors the patient's health automatically.

Smart sensors, a device as IoT, are connected to a patient to collect the patient's heart rate, blood pressure, and electrocardiogram (ECG) and send those data to the Galileo board. The Galileo board would show patient data on this LCD and connect to the computer by USB cable to check their health status. In addition, the microcontroller inside the Galileo board would upload the patient data to a web server via the Internet like Wi-Fi and Ethernet. Thus, the doctor would know the complete medical information of the patient by logging in on the webserver, and the patient can also track their health status on the webserver. Moreover, alert messages would be sent to doctors when the current status is in an emergency. Raw data of patient health status would also be sent to the database server for data analysis.

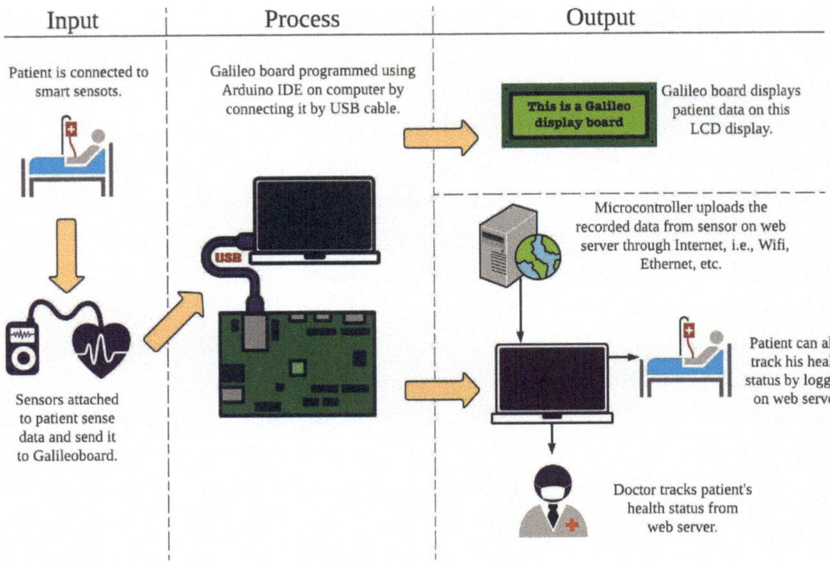

Fig. 4.5 System Architecture of smart healthcare (Gupta et al., 2016).

With the historical data and real-time data, medical experts would provide better medical decisions, and it may result in more life saving and improving patient's health.

It is suggested that the above model should combine with cloud computing and big data analysis for a few reasons. As discussed before, cloud computing providers allow users to choose pay-per-use-basis with the flexibility of selecting appropriate cloud computing models. In this case, medical experts and patients would access the cloud server when they want or in an emergency. Thus, the cost of data storing would be lower compared to the traditional database. Furthermore, as a hospital contains massive data of current and previous patients, there would be a lot of data in heterogeneous formats, which would be considered big data. By applying big data analysis, medical experts would be able to deal with those complex data to conduct any types of data analytics they want. Hence, medical experts can conduct various functional analyses to help patients with the help of extensive data analysis. As a result, by integrating cloud computing and big data analytics, the functions of smart healthcare with IoT would be enhanced. More lives would be saved, and more patients' health would be improved.

Fig. 4.6 reveals a proposed model of smart traffic (Javaid et al., 2018). This model is designed for controlling the road networks and would be divided into three layers:

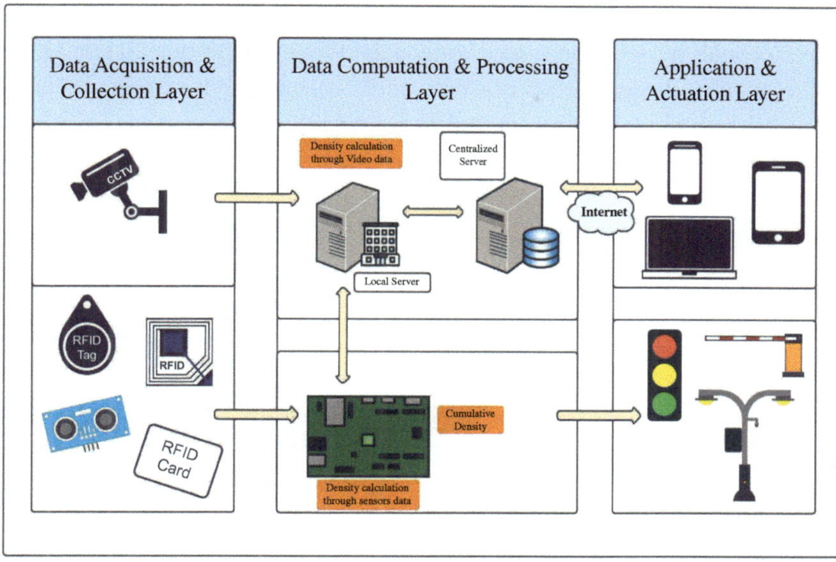

Fig. 4.6 System Architecture of smart traffic (Javaid et al., 2018).

(1) Data Acquisition and Collection layer
(2) Data Processing and Decision-making layer
(3) Application and Actuation layer

Data Acquisition and Collection layer include surveillance cameras, ultrasonic sensors, RFIDs, smoke sensors, and flame sensors for traffic detection. For the cameras, detecting the road traffic for effective and easy maintenance are the functions of surveillance cameras. Video streams captured by surveillance cameras would be sent to a local server, and some detection algorithm would be used to reduce noise in that video stream. Then some microcontroller would receive the density measured through image processing by a local server. For the sensors, increasing the accuracy of the system is the function of ultrasonic sensors. Also, sensors are an integral part of detecting traffic density and measuring distance by sending out a sound wave of a specific frequency and listening for that sound wave to bounce back. The Data Processing and Decision-making layer decide management policies depending on road traffic conditions. The system would calculate traffic density. With the ability to know traffic density, the system would choose the suitable traffic light for practical situations. Thus, the system helps to solve the problem of rush traffic roads. In addition, the regular operation would be stopped when emergency vehicles are detected on the roads. A green signal would be given until that emergency vehicle passes to that intersection. The application and Actuation layer includes two types of information delivered. The first one is the duration of a green signal from node to traffic signal. The second one is daily, weekly, monthly and yearly reports to the administration of smart traffic management systems by using the website applications in a centralized server. In detail, the system would calculate and save the data of rush interval, daily, weekly, monthly and yearly reports. The above data would be generated and displayed on the web application linked to a centralized server. Moreover, emergencies would also be detected by sensors, and alert messages would be sent to the centralized server and respective departments. When the centralized server and respective department know the situations by mobile application, they will conduct further actions to deal with the emergency.

It is recommended that the above model should combine with cloud computing and extensive data analysis for various reasons. In this case, the cost of data storing would be lower compared to the traditional database. Users of the respective department only access the cloud server for emergencies or viewing data for special needs. Furthermore, as this system includes many current and previous traffic data, there would be a lot of data in heterogeneous formats, which would be considered big data. By applying big data analysis, users would be able to deal with those complex data to conduct any data analytics types. Hence, experts can conduct various useful analyses to lower the risk of rush traffic with the help of big data analysis. As a result, by integrating cloud computing and big data analytics, the functions of smart traffic with IoT would be enhanced. Hence, transportation in cities would be more efficient and effective, and accidents on traffic roads would be better controlled and monitored.

Although applying IoT, cloud computing, and big data collective intelligence enhances smart cities' development and operation and would benefit citizens. There would be some challenges in adopting IoT, cloud computing, and big data in smart cities that should be carefully addressed to utilize the advantages fully.

4.5 Challenges for adopting IoT, cloud computing, and big data collective intelligence in a smart city

The possible advantages of applying IoT, cloud computing, and big data collective intelligence in smart cities have been discussed in the above sections. But, the following obstacles should be tackled for practical application in IoT, cloud computing, and big data collective intelligence in smart cities:

(1) Data privacy issue

The IoT environment involves many kinds of technologies and communication standards, but a unified standard policy regarding security and privacy requirements does not exist (Alaba et al., 2017). Moreover, although big data captured by IoT in smart cities is suggested to be stored in cloud databases, some may not agree or be able to adopt cloud computing and adopt traditional databases. As that big data would contain many kinds of personal information, hackers would like to access that data for illegal actions. For instance, when hackers access smart health systems and know patients' names, addresses, phone numbers, or email addresses, those patients' identification would be easily thieved by hackers. Thus, guaranteeing data authenticity and confidentiality is important to protect people's privacy in the environment of smart cities.

(2) Design and implementation issue

As the IoT environment involves many kinds of High-Tech devices, the cost of devices purchased during the design stage would be high. Moreover, talents are needed for the design of smart cities to avoid the possibility of maintenance or redesign. Therefore, the cost of maintenance or redesign would be lower, enhancing the chance of real-world implementation of smart cities and reducing resources wasted. Government support for technologists and education may contribute a lot to the development of smart cities. So a part of smart cities' maintenance or redesign cost would be prevented. In addition, during the implementation stage, it is recommended to provide more education to citizens as the application of smart cities would be new to citizens. As a result, some may refuse to adopt smart cities if they do not have enough information about them, and it may enlarge barriers to building up smart cities.

(3) Data issue

Data captured by IoT must be accurate and reliable for extensive data analysis to conduct the right decision in every application of smart cities. Every mistake or error in data would highly affect the predictions and analysis. As a result, people's living quality would not improve or even decrease via adopting smart cities. Therefore, we should make sure that the data collected by IoT devices are error-free.

(4) Legal issue

As big data in smart cities may contain much sensitive information in different databases or cloud databases. The government should help with data protection and security. That big data includes massive data at the country level such as name, address, health condition, and phone number for all citizens, including celebrities. Hackers would spend much effort in unauthorized access to those big data because those data are valuable for conducting cyber crimes. So, it is suggested that the government should ensure that big data is secure to prevent any serious criminals.

References

Al Nuaimi, E., Al Neyadi, H., Mohamed, N., & Al-Jaroodi, J. (2015). Applications of big data to smart cities. *Journal of Internet Services and Applications*, *6*(1), 1–15.

Alaba, F. A., Othman, M., Hashem, I. A. T., & Alotaibi, F. (2017). Internet of things security: A survey. *Journal of Network and Computer Applications*, *88*, 10–28.

Badawi, W. A. (2019). Underground pipeline water leakage monitoring based on IOT. *International Journal of MC Square Scientific Research*, *11*(3), 01–08.

Benevolo, C., Dameri, R. P., & D'Auria, B. (2016). Smart mobility in smart city. In *Empowering organisations* (pp. 13–28). Cham: Springer.

Calavia, L., Baladrón, C., Aguiar, J. M., Carro, B., & Sánchez-Esguevillas, A. (2012). A semantic autonomous video surveillance system for dense camera networks in smart cities. *Sensors*, *12*(8), 10407–10429. https://doi.org/10.3390/s120810407.

Chen, X., Li, C., Tang, Y., & Xiao, Q. (2018). An internet of things based energy efficiency monitoring and management system for machining workshop. *Journal of Cleaner Production*, *199*, 957–968.

Choi, C. S., Jeong, J. D., Lee, I. W., & Park, W. K. (2018 January). LoRa based renewable energy monitoring system with open IoT platform. In *2018 International conference on electronics, information, and communication (ICEIC)* (pp. 1–2). IEEE.

Dahdouh, K., Dakkak, A., Oughdir, L., & Abdelali, I. (2020). Improving online education using big data technologies. In F. Altlnay (Ed.), *The role of technology in education*. London: IntechOpen. https://doi.org/10.5772/intechopen.88463.

Damen, D., & Hogg, D. (2012). Detecting carried objects from sequences of walking pedestrians. *IEEE Transactions on Pattern Analysis and Machine Intelligence*, *34*, 1056–1067.

Dhingra, S., Madda, R. B., Gandomi, A. H., Patan, R., & Daneshmand, M. (2019). Internet of things mobile–air pollution monitoring system (IoT-Mobair). *IEEE Internet of Things Journal*, *6*(3), 5577–5584.

El Zouka, H. A., & Hosni, M. M. (2021). Secure IoT communications for smart healthcare monitoring system. *Internet of Things*, *13*, 100036. https://doi.org/10.1016/j.iot.2019.01.003.

Furht, B., & Villanustre, F. (2016). Introduction to big data. In *Big data technologies and applications* (pp. 3–11). Cham: Springer.

Gupta, P., Agrawal, D., Chhabra, J., & Dhir, P. K. (2016 March). IoT based smart healthcare kit. In *2016 International conference on computational techniques in information and communication technologies (ICCTICT)* (pp. 237–242). IEEE.

Hossain, M. Y., & George, F. P. (2018 October). IOT based real-time drowsy driving detection system for the prevention of road accidents. In *Vol. 3. 2018 International conference on intelligent informatics and biomedical sciences (ICIIBMS)* (pp. 190–195). IEEE.

Iqbal, R., Doctor, F., More, B., Mahmud, S., & Yousuf, U. (2020). Big data analytics: Computational intelligence techniques and application areas. *Technological Forecasting and Social Change*, *153*, 119253.

Javaid, S., Sufian, A., Pervaiz, S., & Tanveer, M. (2018 February). Smart traffic management system using internet of things. In *2018 20th International conference on advanced communication technology (ICACT)* (pp. 393–398). IEEE.

Kaur, M. J., & Maheshwari, P. (2016, March). Building smart cities applications using IoT and cloud-based architectures. In *2016 International conference on industrial informatics and computer systems (CIICS)* (pp. 1–5). IEEE.

Kim, B., Kim, M., & Jeong, J. (2020 July). Estimation of greenhouse gas emissions in cement manufacturing process through blockchain and SSL based IoT data analysis. In *International conference on computational science and its applications* (pp. 634–645). Cham: Springer.

Kim, J., & Park, N. (2019). Lightweight knowledge-based authentication model for intelligent closed circuit television in mobile personal computing. *Personal and Ubiquitous Computing*, 1–9. https://doi.org/10.1007/s00779-019-01299-w.

Kodali, R. K., & Mandal, S. (2016 December). IoT based weather station. In *2016 International conference on control, instrumentation, communication and computational technologies (ICCICCT)* (pp. 680–683). IEEE.

Lookmuang, R., Nambut, K., & Usanavasin, S. (2018 May). Smart parking using IoT technology. In *2018 5th international conference on business and industrial research (ICBIR)* (pp. 1–6). IEEE.

Madakam, S., Ramaswamy, R., & Tripathi, S. (2015). Internet of Things (IoT): A literature review. *Journal of Computer and Communications*, *3*, 164–173.

Malche, T., & Maheshwary, P. (2017 February). Internet of things (IoT) for building smart home system. In *2017 International conference on I-SMAC (IoT in social, mobile, analytics and cloud)(I-SMAC)* (pp. 65–70). IEEE.

Muntjir, M., Rahul, M., & Alhumyani, H. A. (2017). An analysis of internet of things (IoT): Novel architectures, modern applications, security aspects and future scope with latest case studies. *International Journal of Engineering Research & Technology*, *6*(6), 422–447.

Narendran, S., Pradeep, P., & Ramesh, M. V. (2017 October). An internet of things (IoT) based sustainable water management. In *2017 IEEE global humanitarian technology conference (GHTC)* (pp. 1–6). IEEE.

Niyato, D., Hossain, E., & Camorlinga, S. (2009). Remote patient monitoring service using heterogeneouswireless access networks: Architecture and optimisation. *IEEE Journal on Selected Areas in Communications*, *27*(4), 412–423.

Oussous, A., Benjelloun, F. Z., Lahcen, A. A., & Belfkih, S. (2018). Big data technologies: A survey. *Journal of King Saud University-Computer and Information Sciences*, *30*(4), 431–448.

Pappu, S., Vudatha, P., Niharika, A. V., Karthick, T., & Sankaranarayanan, S. (2017). Intelligent IoT based water quality monitoring system. *International Journal of Applied Engineering Research*, *12*(16), 5447–5454.

Puthal, D., Sahoo, B. P. S., Mishra, S., & Swain, S. (2015). Cloud computing features, issues and challenges: A big picture. In *2015 International Conference on Computational Intelligence and Networks, 12–13 Jan 2015, Odisha, India*IEEE.

Rajaraman, V. (2016). Big data analytics. *Resonance*, *21*(8), 695–716.

Rose, K., Eldridge, S., & Chapin, L. (2015). *The internet of things: An overview. Vol. 80* (pp. 1–50). The Internet Society (ISOC).

Saifuzzaman, M., Moon, N. N., & Nur, F. N. (2017 December). IoT based street lighting and traffic management system. In *2017 IEEE region 10 humanitarian technology conference (R10-HTC)* (pp. 121–124). IEEE.

Saraju, P. M., Choppali, U., & Kougianos, E. (2016). *Everything you wanted to know about smart cities*. Retrieved from https://www.researchgate.net/publication/306046857_Everything_You_Wanted_to_Know_About_Smart_Cities.

Segura Garcia, J., Pérez Solano, J. J., Cobos Serrano, M., Navarro Camba, E. A., Felici Castell, S., Soriano Asensi, A., & Montes Suay, F. (2016). Spatial statistical analysis of urban noise data from a WASN gathered by an IoT system: Application to a small city. *Applied Sciences*, *6*(12), 380.

Strømmen-Bakhtiar, A., & Razavi, A. R. (2011). Cloud computing business models. In Z. Mahmood, & R. Hill (Eds.), *Chapter 3. Cloud computing for enterprise architectures.*

Computer communications and networks (pp. 43–60). London: Springer. https://link.springer.com/chapter/10.1007/978-1-4471-2236-4_3.

Talari, S., Shafie-khah, M., Siano, P., Loia, V., Tommasetti, A., & Catalão, J. (2017). A review of smart cities based on the internet of things concept. *Energies, 10*(4), 421. https://doi.org/10.3390/en10040421.

Wang, J., & Zhang, J. (2016). Robust object tracking in infrared video via adaptive weighted patches. *Mathematical and Computational Applications, 22*(1), 3. https://doi.org/10.3390/mca22010003.

Yaqoob, I., Hashem, I. A. T., Gani, A., Mokhtar, S., Ahmed, E., Anuar, N. B., & Vasilakos, A. V. (2016). Big data: From beginning to future. *International Journal of Information Management, 36*(6), 1231–1247.

CHAPTER FIVE

Smart energy network

The current carbon-based energy network is undergoing changes and revolutions driven by the ever-growing concerns on carbon emissions and air pollutions and the safety and convenience of energy consumption. This global evolutionary transformation is toward a smart energy network as it is reinforced by the emergence of low-cost bandwidth and the rapid development of intelligent energy management technologies. The smart energy network can release the growing burden of energy costs due to increased energy efficiency and shaving the bills (Liu et al., 2019). In addition, the smart energy network will suffer fewer blackouts and brownouts with the enhancement capacity to recover from failures and breakdowns.

In general, the smart energy network is an integrated part of the whole smart city Internet of Things (IoT) solution, and it can bring a comfortable, convenient, safe, and energy-conserved environment to households and buildings (Gaur et al., 2015; Saad Al-Sumaiti et al., 2014). China is one of the world's largest markets in power transmission and distribution and, with the government's determination to significantly reduce the percentage of carbon intensity for its GDP growth. She has experienced a remarkable development in the smart energy network for the aspects such as equipment, communication device, and integrated solution. In this chapter, the smart energy network in China will be discussed and illustrated, and an actual product designed for the smart energy network will be introduced as an example.

Since 2010, smart white goods, electricity meters, water meters, gas meters, heat meters, home/building gateways, and automatic meter reading systems have been promoted by the major manufacturers in China. This widespread trend is an essential step in developing a low-carbon and circular economy in China. At the same time, the IoT concept has been recognized by the government and experts since 2009. It has been risen to the level of national strategy and promoted to an emerging industry. The country attaches great importance to the "12th Five-Year Plan," which clearly proposed that the next generation of national infrastructure for broadband convergence and security should be developed to promote IoT applications (Chen et al., 2014). As a specific IoT application of smart homes and

Collective Intelligence for Smart Cities
https://doi.org/10.1016/B978-0-12-820139-8.00007-3

buildings, the smart city energy network should be user-centric and handle the daily business on the energy demand side. Thus, the smart city energy network could realize its ability of intelligent operations with the IoT application (Sahani et al., 2017).

In the future, the smart city's energy demand-side network needs to meet the need for energy management and the intelligent networks of homes and buildings, bringing comfort, convenience, safety, energy conservation, and environmental protection to household users (Zhang et al., 2015). Its market prospects are extensive. Smart city energy network includes smart household energy network and smart building energy network. The former is for home users, and the latter is for corporation users—the main difference between the two lies in the design of the IoT application layer. The former is a part of the IoT application software for smart home gateways, and the latter is a part of the IoT application software for smart buildings.

The architecture of the energy network in smart cities includes core equipment and systems. The core equipment is a gateway for homes and buildings. The home gateway provides an interactive platform for energy, medical insurance, transportation, and life information to realize the "interactive characteristics" of the smart energy network. At the same time, the role of the home gateway also includes the management of concentrators, smart household appliances, and smart terminals like water, electricity, gas, and heat meters. The company gateway generally only provides an energy collection and interaction platform to manage the company's intelligent terminals on water, electricity, gas, and heat meters. The core systems are energy network IoT cloud platforms and enterprise-level applications. The energy network in smart cities enables interactions between people and things, things and things, things, and people, to perform information exchange and remote access and control on information.

The smart city energy network platform is an essential part of the smart city IoT overall solution, and it is a complete household IoT cloud application system. Its core content is shown in Fig. 5.1.

The energy network in smart cities can promote the transformation of energy demand-side management and user's home experience, support the national development strategy, establish a modern service industry market for supervision on urban energy demand-side information, and build a "conservation-minded society." Due to many end-family and company users, the energy network platform will be one of the critical IoT application platforms in China's IoT architecture that implements as a business model. On this basis, China's fourth-largest operator, the IoT operator, will be born.

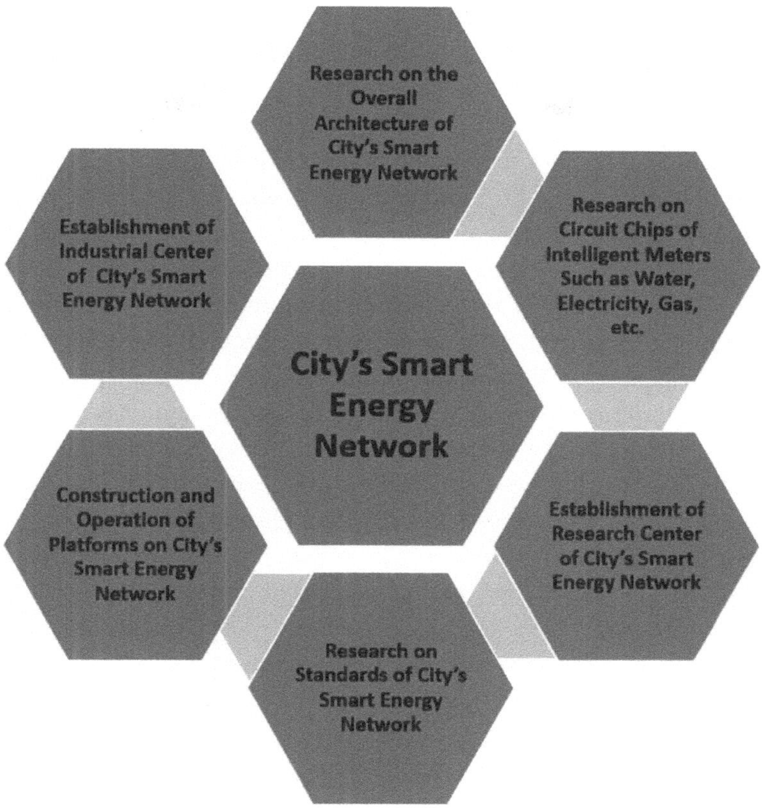

Fig. 5.1 The core content of the energy demand-side network for smart cities.

Smart household energy networks can extend to the smart building energy network. The smart building energy network pays more attention to interface expressions such as charts at the presentation level. The following sections of this chapter focus on the smart household energy network.

5.1 Smart home

The field of smart homes is an important area for IoT applications in family lives (Samuel, 2016). The establishment of an intelligent and well-off demonstration community has become a key development direction in China. "Smart home system" is to connect water, electricity, gas, lighting, audiovisual, security, communication, temperature control, and other equipment in the household through the household domain network to achieve a better working synergy. Residents can remotely enter their house through the terminal control to monitor and alarm house conditions, visitors, accidents, etc. Meanwhile, they can also remotely control the home appliances, thereby improving the quality of life (Fan et al., 2017; Kadam et al., 2015).

The demand-side information on household energy, as well as the loss of resources, is difficult to be monitored and managed. Thus, China's economic development and demand-side management of water, electricity, gas, heat, and other resources in people's daily lives have been plagued by these problems for a long time. There are numerous cities in China, and the number of urban populations is huge. In the process of building a "saving city," we will promote the marketization of information management operations on the demand-side of water, electricity, gas, and heat resources in Chinese cities. Thus, the operation platforms and service markets on water, electricity, gas, and heat resource in cities will be established. This establishment will be one of the most critical IoT application platforms in China's business IoT architecture model implementation. At present, all regions of China have mature network IP backbone networks, TD-SCDMA, and other network layer support conditions. They mainly develop the energy demand-side IoT access platforms and produce IoT equipment of the information terminal system.

In the construction of the IoT, the research and development of the IoT on the household energy demand-side is the core of the smart home system. The asymmetric power carrier communication device in the system controls the family's white household appliances and security facilities simultaneously. In other words, the IoT's asymmetric power carrier communication device on the home energy demand-side is also the core of the smart grid in households. A smart home system is a set of intelligent systems based on houses and integrated with intelligent facilities such as network communications, intelligent buildings, and smart appliances to provide a convenient and comfortable household environment. The smart home IoT platform can provide users with an efficient, safe, convenient, and environmentally friendly living environment. Intelligent interactive terminals, intelligent meters for electricity, water, heat, and gas, and smart home appliances are indispensable to realize intelligent home systems (Arif et al., 2013; Carratù et al., 2018).

From the perspectives of industry demands, the existing energy demand-side systems still have significant deficiencies, mainly in the following aspects:

(1) Focusing on service providers: Although most of the provinces in China have opened payment systems for water, electricity, heat, and gas, the scale of these systems is generally small, and many cities manage these systems by the district. It is challenging to realize the intelligence of the systems and improve the satisfaction of residents.

(2) Information silos: At present, the energy demand-side systems in China are constructed, operated, and managed independently. The construction of the city's IoT includes many aspects, such as intelligent transportation, intelligent medical treatment, and intelligent fire protection. IoT for the energy demand-side is an essential part of the overall smart city and the cornerstone of the birth of China's fourth-largest IoT operator. The smart energy network needs to integrate into the whole smart city network to work.

(3) Lack of IoT intentions: Due to technical limitations, the existing energy demand-side systems are implemented per community. Thus, it lacks overall plans and measures at a city level, failing to enable user interaction and unable to store and retrieve city-level cloud data.

(4) Insufficient service capabilities: At present, the energy-demand-side system intelligently provides traditional monitoring functions, such as timely alarms and effective handling of payment on water, electricity, heat, and gas. However, it is not compelling enough regarding user satisfaction improvement, operation performance enhancement of facilities like water, electricity, and gas, and energy department's capabilities improvement on dynamic management and proper supervision for the city's energy consumption. Therefore, although the system is implemented, the functions of IoT cannot be realized in communities.

The smart energy IoT network on the energy demand-side can solve user adoption problems and regulation issues from the supervision departments and create its expected economic and social value.

5.2 Status of China and foreign countries

The IoT for smart homes is recognized as the main component of the IoT, and it is the IoT project that is most closely related to residents and has the most investment value. Smart homes provide convenience for people's living and transportation and make the equipment and appliances at home operate automatically according to the habits and status of users, thereby improving people's quality of life. According to the estimation from Frost & Sullivan, a world-renowned consulting firm, China's smart home/building output value will be at least 1 trillion yuan in 2020.

According to the relevant data, the total consumption of people with an annual income level of more than 56,000 yuan can reach about 17% of the total purchasing power. Based on this calculation, the total market of the system for household safety in this group is more than 14 million units: the prospect of this consumption is huge. The main functions of the new generation of intelligent interactive terminals include home energy management, home affairs management, indoor environment, and user information collection, analysis and transmission, 3G remote video monitoring, distributed energy power information collection and management, smart home appliance management, charge and discharge management of electric vehicles, security and access control, public information release, multimedia entertainment, video message memo, automatic Internet access of electrical equipment, advertisement release, automatic alarm and so on. Fig. 5.2 shows the main functions of the intelligent interactive terminals.

At present, there are a large number of new technologies and products related to smart homes in the market, including

(1) Intelligent interactive terminals
(2) Smart sockets
(3) Intelligent lighting switches
(4) Wireless sensor modules
(5) Load control modules
(6) Wireless meter reading and communication module for meters of water, electricity, gas, and heat
(7) Smart home software development modules on mobile phone side
(8) Home intelligent network server

Among these products, intelligent interactive terminals are produced by domestic and foreign companies such as Haier and Midea of China, Aituli of France, Elster of the United Kingdom, and Shenshus of the United States. Producers of smart meters of water, electricity, gas, and heat include Hangzhou Jingda Electronics, Hangzhou Shuibiao Limited, Suntront Technology, Texas Instruments, and many other companies.

Currently, the construction of the IoT on the energy demand side is still in its infancy, and the cloud platforms on the energy demand side have not yet appeared on the market.

5.3 Market analysis and social value

The IoT has become a scientific research subject strongly supported by the Chinese government. Some wireless sensor devices with security and automatic detection functions have been applied in many aspects. Smart homes with the IoT concept have also been highly publicized, stimulating

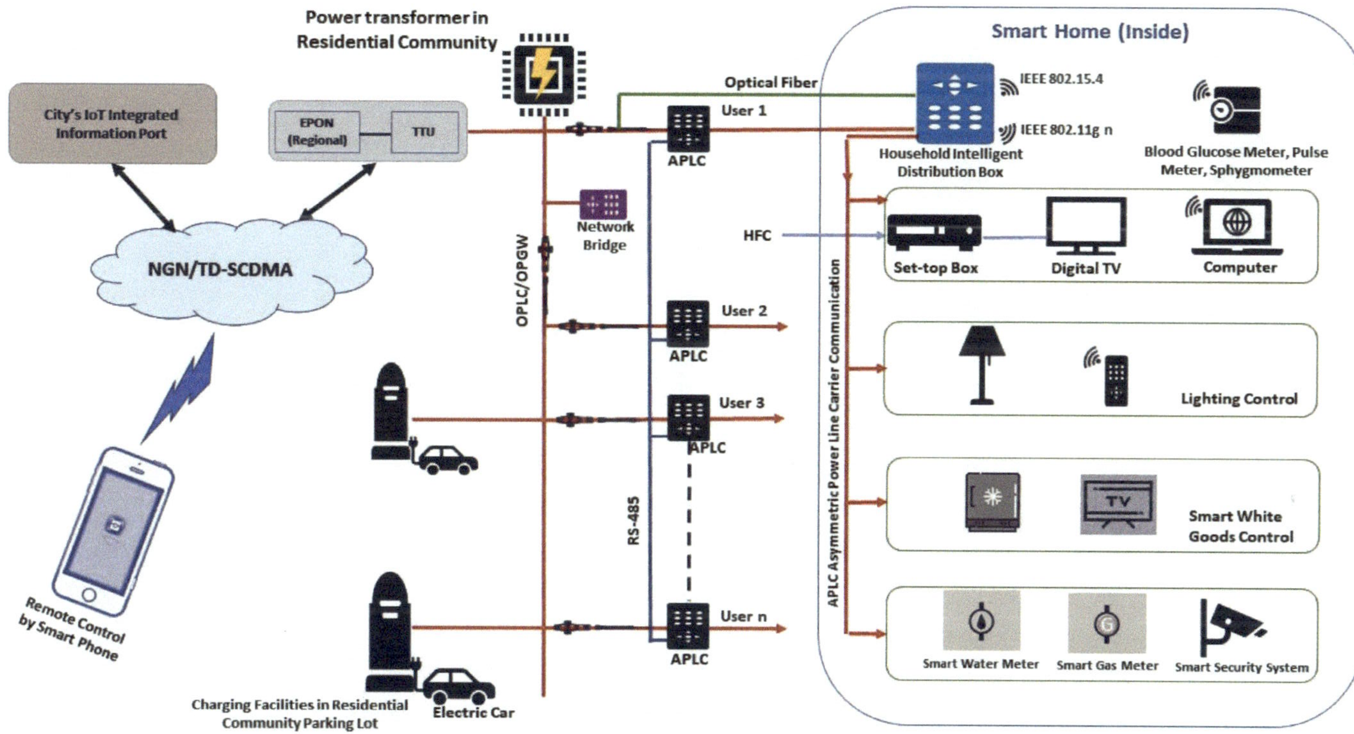

Fig. 5.2 Main functions of the new generation of the intelligent interactive terminals.

the rapid growth of the smart home industry in China. In the next few years, the smart home industry will be rapidly developed.

At the same time, China's support for green buildings is also very strong. In the smart equipment required in this regard, a new market of 200 billion to 300 billion yuan has been opened. Intelligent interactive terminals that integrate home energy management functions and intelligent meters of water, electricity, gas, and heat are an essential part of intelligent green buildings. Their market will be promoted due to the promotion of intelligent green buildings. Real estate competition is becoming increasingly fierce. To cater to consumers' preferences on environmental protection and smart homes, many real estate developers will also actively adopt smart home systems to attract consumers. The market space then becomes very broad.

With the popularity of smart communities, the market demand for smart home systems will also be extensive. According to the survey's data from the relevant agencies, with improved people's living standards and consumption habits, the demand for a stylish, comfortable, and energy-saving home environment is increasing dramatically. Since 2010, over 57 million middle-class households in China, about 60% of mid-to-high-end consumers, are willing to install smart home systems. If 60% of mid-to-high-end households are willing to install smart home systems, it means it will create a huge market of hundreds of billions of yuan.

The social value of the IoT for smart homes includes the following:

(1) Focusing on users, improving people's life quality and user satisfaction.

(2) Through the construction of the IoT for smart homes, the power expenditure in a household will be reduced, and due contributions will be made to China's development of the green, sustainable and low-carbon economy model and the development of smart cities.

(3) Employment promotion. The construction of the IoT for smart homes is a comprehensive project that requires talents in various fields such as sensors, control systems, networks, software, etc. In addition to research and development, it will also involve producing smart home equipment, monitoring operations of the network, and other value-added services. A large number of employment opportunities will be increased. Driven by "Industry-University-Research Collaboration," more than 700 universities and colleges have applied for the "Internet of Things Engineering" program. According to the statistics, the market size of the IoT industry is close to 200 billion yuan in 2010, which has been a huge driving effect on the employment market.

(4) Promoting innovation and self-owned intellectual property products. As stated in the "Twelfth Five-Year Plan for the Development of the Internet of Things," "the role of enterprises should be brought into full play, and the existing research results of universities and research institutes should be actively used in the field of information perception technology, and information processing technology to catch up with the advanced level in the world, to reach the advanced level in the field of information transmission technology, to enhance information security guarantee capabilities, and strive to break through key core technologies as soon as possible to form a complete Internet of Things technology system and achieve industrialization."

5.4 Overall architecture of system design

The energy demand-side IoT system is an essential part of the comprehensive information platform of the city's IoT. This system enables management departments to directly monitor water, electricity, gas, and heat resources and enables household users to monitor home white goods and energy consumption through wireless networks. The smart city IoT architecture is shown in Fig. 5.3.

From the operational point of view, the IoT for the city's energy demand-side is divided into three levels: the IoT awareness layer, network layer, and management service application layer. The energy demand-side IoT operation platform is shown in Fig. 5.4.

5.5 Smart energy network solutions

The smart energy demand-side IoT solution enables the monitoring and managing of meter terminals for water, electricity, gas, and heat through the control system, which mainly involves four parts.

(1) Household IoT platform for energy-consuming equipment such as water, electricity, gas, heat resources, and white goods.
(2) Enterprise distribution IoT platform for the water supply company, gas company, and heat & power company.
(3) IoT platform for electronic payment and settlement.
(4) IoT platform connecting enterprises and families.

Fig. 5.5 shows a schematic diagram of the energy demand-side IoT access platform.

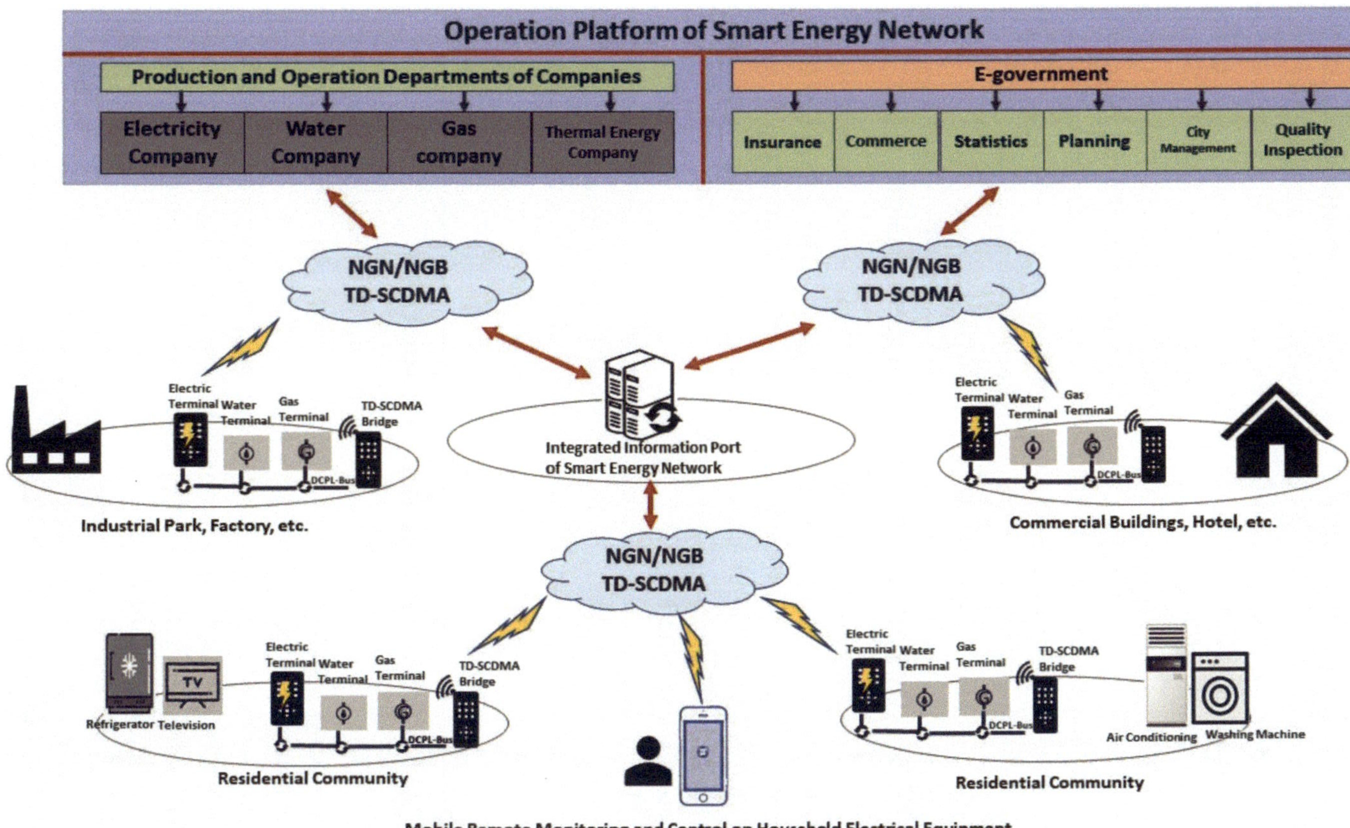

Fig. 5.3 The smart city IoT architecture.

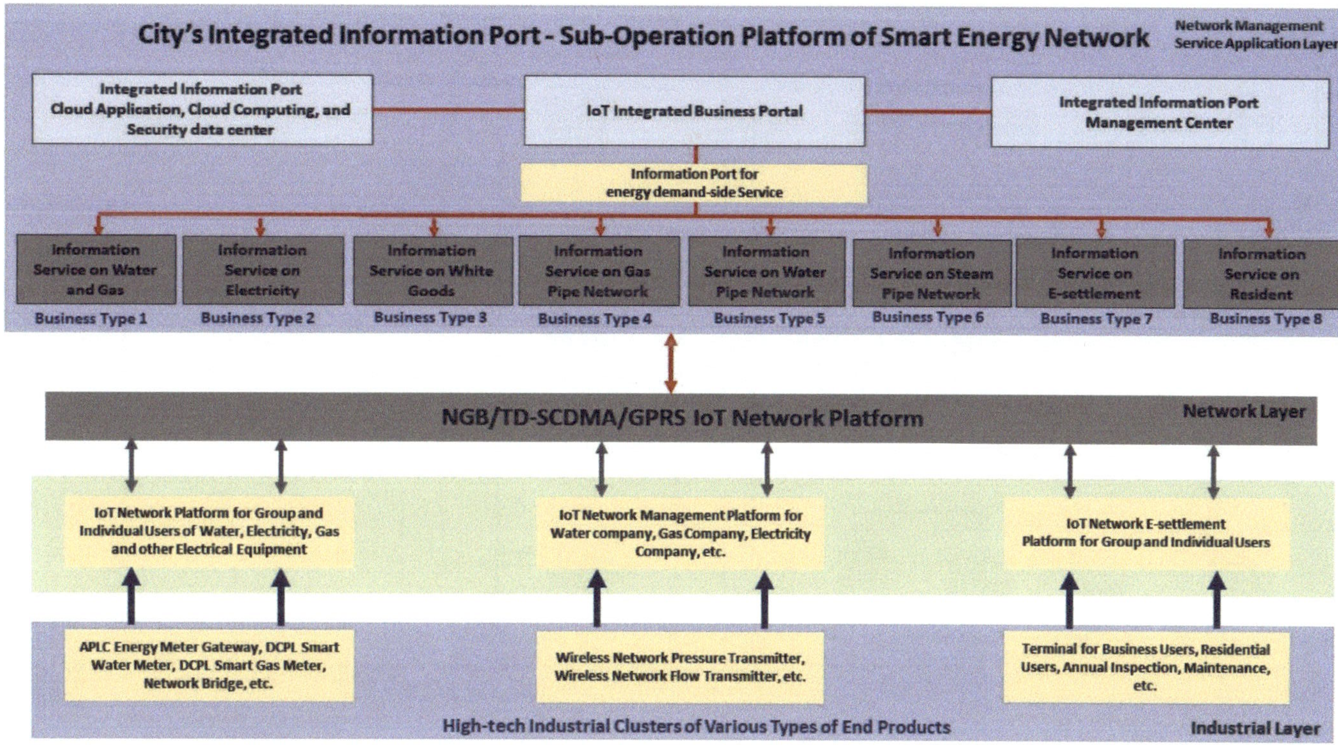

Fig. 5.4 The energy demand-side IoT operation platform.

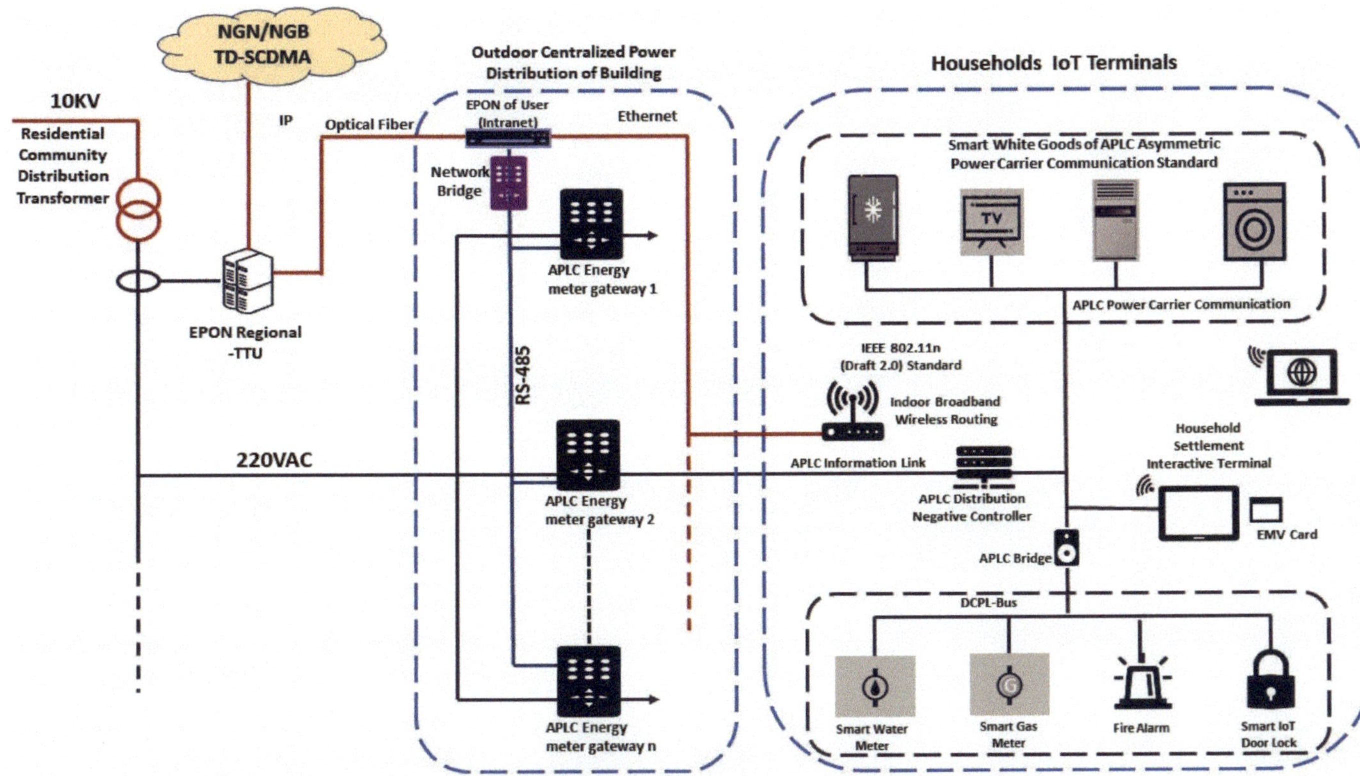

Fig. 5.5 Schematic diagram of the energy demand-side IoT access platform.

5.6 IoT network layer and service management application layer of smart energy network

The network layer is the channel for data communication between the smart energy network control system and the IoT monitoring center in the building. It can use multiple network environments such as wired and wireless, and it can also use multiple operators' networks. The network layer is required to support at least GPRS, WCDMA and TD mobile networks, PSTN and ADSL, and support the Internet exit of existing LANs to transmit data.

The service management application layer is based on the city's public network. It uses modern high-tech methods to remotely monitor the real-time operation of the energy demand-side IoT distributed throughout the city. The energy demand-side IoT operation platform includes the cloud application of the IoT integrated information port, the cloud computing center, cloud security data center, the city's IoT integrated business portal, and the comprehensive management center for IoT integrated information. These components together constitute the energy demand-side service information port. The software of the operation management center of the service management application layer should include a series of modules. At the same time, smart terminals help users perform operations such as querying and receiving targeted advertisements.

5.7 Case example

As energy, especially electricity, practical usage, and safety are being taken seriously by the smart energy network, a Smart Power Extension Bar (SPEB) has been developed to address the issues. The characteristics and architecture of the SPEB will be introduced.

SPEB has been developed as a power extension bar with a proactive monitoring function and a very practical example of an IoT-based product for homes and buildings. Fig. 5.6 shows the system structure of the SPEB.

Fig. 5.7 shows a trendy design of the SPEB. The design is based on the traditional extension unit or power strip connecting the main socket to some electrical appliances and electronic devices. It contains six 13A three-pin plug outlets with four USB power ports.

The SPEB uses a relay control circuit to isolate high voltage levels on the printed circuit board. With an individual ON/OFF button and a high

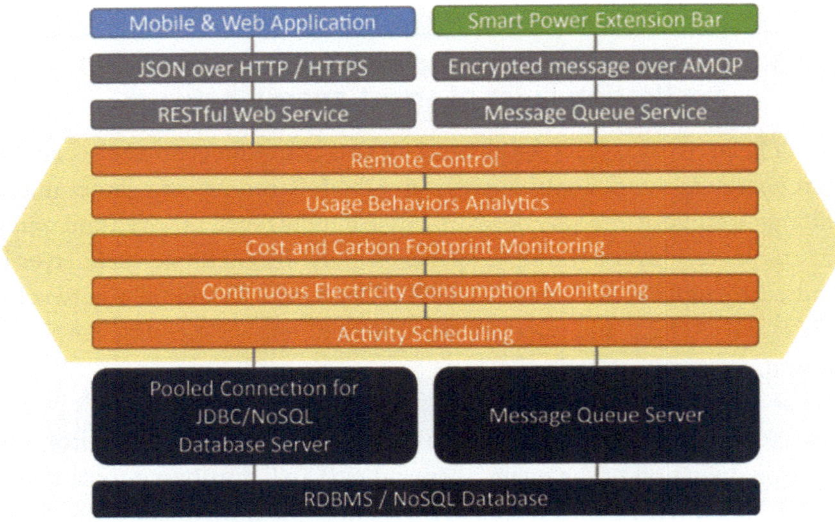

Fig. 5.6 A system structure diagram of SPEB.

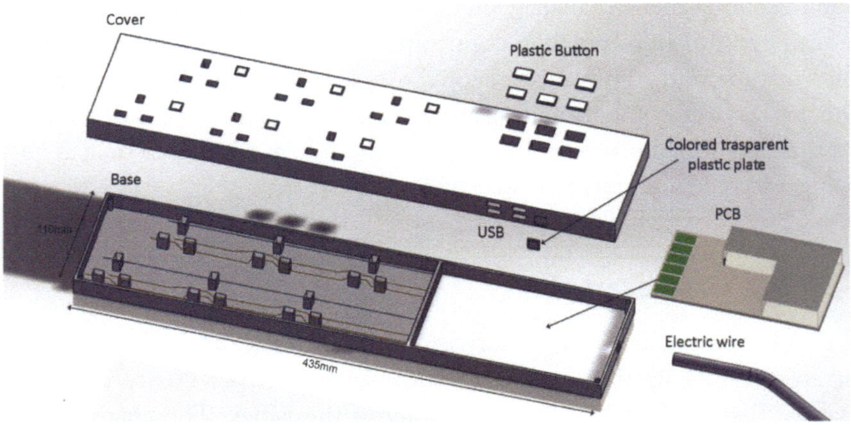

Fig. 5.7 A schematic assembly drawing of SPEB.

accuracy current sensor embedded, each outlet can also be controlled remotely through the internet connection. The SPEB has an indicating light of surge protection enabled. It is essential to protect the electrical appliances and electronic devices connected to the SPEB when an incoming voltage suddenly rises, mainly caused by lightning strikes.

All socket outlets and USB power ports of the SPEB can be automatically turned on or off based on users' preset schedules, and users can also reschedule via an Android-based mobile App when necessary. However, each ON/OFF button (physical buttons) on the extension bar is still the ultimate control of each outlet over the commands received from the App. The SPEB is Wi-Fi-enabled to connect to the internet directly without any middleware connection. The Wi-Fi network should be with Wi-Fi Protected Setup (WPS) protocol. Advanced Message Queuing Protocol (AMQP) is used as a lightweight IoT communication protocol. The SPEB can be connected with the App so that users can monitor each outlet conditions, including electric current, voltage, and energy consumption via the App.

A SPEB is first connected to a power socket on a wall. Other electrical appliances or electronic devices are connected to the SPEB to manage and monitor them via the App on their Android devices. With the SPEB, users can set schedules, monitor energy usage on all electrical appliances and electronic devices, and quickly identify those devices with higher usage frequency and usage distribution. Armed with such continuous data, users can start saving their home's domestic electricity usage by changing user behaviors.

Unlike a traditional analog or digital timer switch with manual settings required, users can use a SPEB to control any electrical appliance or electronic device anytime and anywhere via internet access. Analog or digital timer switches are practical but not "smart" enough if users' schedules are unstable or unsure when someone is at home. For example, users may want their homes to be nice and toasty when they are home from work, but their return time is different every day. With SPEBs, users can turn on a heater, Hi-Fi, smooth lighting, etc., directly from their Android smartphones when they are on the way home.

One of the more appealing functions of the SPEB is energy use and cost monitoring. Many electrical appliances and electronic devices are always in standby mode even though they are not used nearly all day. The standby mode uses account for up to 10%–15% of the electricity bill on average. The use of SPEB can dramatically cut down the unnecessary standby time. Although some similar smart switches are available in the market, they are not affordable for all to have one at home. Indeed, with the energy-monitoring features, these smart switches promise actual energy and money-saving via usage data and standby mode blocking. However, this would be even more useful if such switches could monitor multiple devices on a "power-strip-like" smart switch, which is why the SPEB has been designed and developed.

Major advantages of the Smart Power Extension Bar (SPEB):

(1) Wireless personal power control and continuous electricity consumption monitoring.

(2) High safety in usage with surge protection and overcurrent sensing protection.

(3) Standby mode blocking by SPEB timer function.

With a dedicated, specific mobile App, namely "TGT Power Widget" (Fig. 5.8), a single SPEB can be only registered by one user account while each user account can manage multiple SPEBs. Users can manage and control the entire SPEBs or outlet by outlet. Also, the auto cut-off limit can be

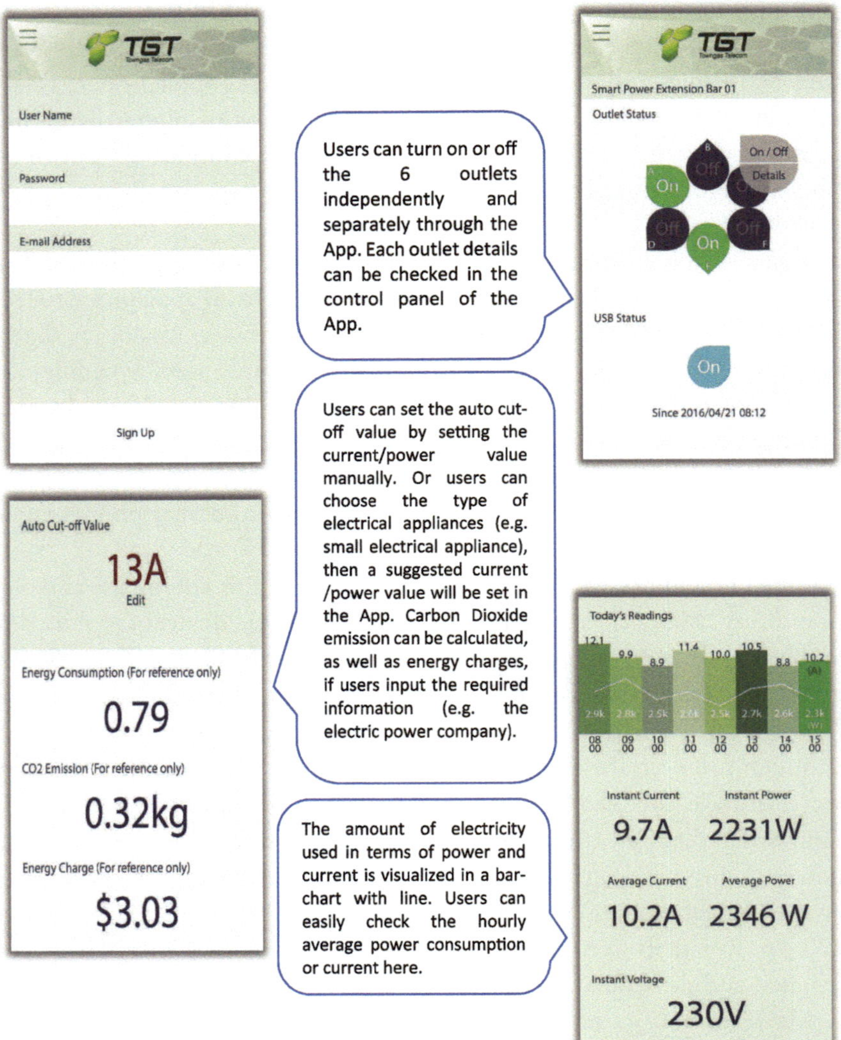

Fig. 5.8 Screen shots of the TGT power widget.

set via the App. Auto cut-off means the power will be shut down when the power/current reaches the predetermined value. By default, if SPEB finds the power consumption sustains an abnormal or ascending trend (e.g. rising toward 13A), an alert will be sent to inform users of immediate action. Except for these alerts for critical situations, users can turn off noncritical notification pop-ups in the App.

References

Arif, A., Al-Hussain, M., Al-Mutairi, N., Al-Ammar, E., Khan, Y., & Malik, N. (2013). Experimental study and design of smart energy meter for the smart grid. In *2013 International renewable and sustainable energy conference (IRSEC)* (pp. 515–520). IEEE.

Carratù, M., Ferro, M., Pietrosanto, A., & Paciello, V. (2018, July). Smart power meter for the IoT. In *2018 IEEE 16th international conference on industrial informatics (INDIN)* (pp. 514–519). IEEE.

Chen, S., Xu, H., Liu, D., Hu, B., & Wang, H. (2014). A vision of IoT: Applications, challenges, and opportunities with China perspective. *IEEE Internet of Things Journal, 1*(4), 349–359.

Fan, X., Qiu, B., Liu, Y., Zhu, H., & Han, B. (2017). Energy visualization for smart home. *Energy Procedia, 105*, 2545–2548.

Gaur, A., Scotney, B., Parr, G., & McClean, S. (2015). Smart city architecture and its applications based on IoT. *Procedia Computer Science, 52*, 1089–1094.

Kadam, R., Mahamuni, P., & Parikh, Y. (2015). Smart home system. *International Journal of Innovative Research in Advanced Engineering (IJIRAE), 2*(1), 81–86.

Liu, Y., Yang, C., Jiang, L., Xie, S., & Zhang, Y. (2019). Intelligent edge computing for IoT-based energy management in smart cities. *IEEE Network, 33*(2), 111–117.

Saad Al-Sumaiti, A., Ahmed, M. H., & Salama, M. M. (2014). Smart home activities: A literature review. *Electric Power Components & Systems, 42*(3–4), 294–305.

Sahani, B., Ravi, T., Tamboli, A., & Pisal, R. (2017). IoT based smart energy meter. *International Research Journal of Engineering and Technology (IRJET), 4*(4), 96–102.

Samuel, S. S. I. (2016, March). A review of connectivity challenges in IoT-smart home. In *2016 3rd MEC international conference on big data and smart city (ICBDSC)* (pp. 1–4). IEEE.

Zhang, X., Chan, S. H., Ho, H. K., Tan, S. C., Li, M., Li, G., Li, J., & Feng, Z. (2015). Towards a smart energy network: The roles of fuel/electrolysis cells and technological perspectives. *International Journal of Hydrogen Energy, 40*(21), 6866–6919.

CHAPTER SIX

Smart firefighting and fire protection

The technologies in the emerging digital era, which includes the Internet of Things (IoT), machine-to-machine communication, Big Data analytics, and so on, has provided the solutions for countries or cities to achieve safety and effectiveness in firefighting and fire protection by the establishment of a smart firefighting system (Cicioğlu & Calhan, 2020; Gaur et al., 2019; Salhi et al., 2019; Wu et al., 2017). The system can promote the reform in a country on various aspects, for example, city fire prevention, rapid disaster response, social security, and emergency plan management (Mahgoub et al., 2020; Ryu, 2015; Sharma et al., 2020; Zhang et al., 2020). It can also support the national harmonious development strategy, establish a citywide fire safety information supervision, and promote the annual inspection of fire prevention units, equipment maintenance, fire safety education. More importantly, it can improve the fire prevention management department and society to integrate and enhance disaster prevention.

The system is based on a framework to (1) collect large quantities of information and data from various sources, (2) perform big data analysis and prediction, (3) disseminate analysis results with relevant information to the concerned parties, such as fire departments, emergency command centers, and residential communities, for the facilitation of decision marking, and (4) tackle the challenges (e.g., environmental hazards) of the fire scene. The system is an integrated platform for the overall IoT solution of a smart city.

The city's smart firefighting IoT platform mainly contains three components: fire monitoring, fire extinguishing, and fire management. In China, the city's smart firefighting IoT platform serves as one of the most important IoT application platforms for nationwide IoT architecture implementation. In the chapter, a smart firefighting IoT platform would be discussed to demonstrate the concept and application of smart firefighting and fire protection.

Collective Intelligence for Smart Cities
https://doi.org/10.1016/B978-0-12-820139-8.00002-4

6.1 Status of China

In March 2006, the Fire Bureau of the Ministry of Public Security of China held a conference on technical demonstration and application promotion for buildings' fire protection facilities and remote monitoring systems for fire alarms in Fuzhou. The conference minutes stated that "we must vigorously promote and standardize the construction and application of remote monitoring systems for fire safety." In October 2007, the Ministry of Construction of China issued the "Technical Specifications for Urban Fire Remote Monitoring System" (GB50440-2007), which specifies the content, performance, system architecture, and system implementation of the construction IoT fire firefighting.

In August 2008, the director of the Fire Bureau of the Ministry of Public Security of China emphasized in a conference on the construction and application of remote monitoring systems for city fire safety. The construction of remote monitoring systems for city fire safety throughout the country is a vital informatization application. This application shows scientific and technological power that is in line with the Party Committee's focus. In September 2008, the Fire Bureau of the Ministry of Public Security formulated the "Guiding Opinions on Promoting and Standardising the Construction and Application of City Fire Safety Remote Monitoring System." The bureau stated that "a remote fire safety monitoring system in the city is an important scientific and technological means to strengthen public fire safety management. It has a very important role in strengthening the fire safety management, moving forward the fire prevention barrier, disposing of fires quickly, and improving the comprehensive ability of cities to prevent and control fires." Also, cities above the prefecture level should establish remote monitoring centers for city fire safety. The towns where the central government is located establish remote monitoring centers as needed.

In July 2011, the National Standardization Management Committee issued Announcement No. 12, approving the release of GB 26875-2011 "City Fire Protection Remote Monitoring System" series of national standards. This series of standards comprehensively regulates all aspects of the construction of the City firefighting remote monitoring system, including six national standards, as shown in Table 6.1.

According to the "Twelfth Five-Year Plan for the Development of the Internet of Things," the existing firefighting network monitoring system still has significant shortcomings from industry needs. The main problems are

Table 6.1 National Standards of GB 26875-2011 "City firefighting Remote Monitoring System."

Item no.	Description
GB 26875.1-2011	City firefighting remote monitoring system Part 1: User information transmission device
GB 26875.2-2011	City firefighting remote monitoring system Part 2: Functional requirements for communication server software
GB/T 26875.3-2011	City firefighting remote monitoring system Part 3: Alarm transmission network communication protocol
GB/T 26875.4-2011	City firefighting remote monitoring system Part 4: Basic data items
GB 26875.5-2011	City firefighting remote monitoring system Part 5: Functional requirements for process management software
GB 26875.6-2011	City firefighting remote monitoring system Part 6: Functional requirements for information management software

similar to the smart city energy network. The shortcomings include the limited IoT-based firefighting system users, independent construction, operation, and management of various monitoring systems, insufficient service capabilities, etc. The development of a city firefighting IoT platform helps solve the existing monitoring problem in the firefighting networks and creates dual economic and social values.

IoT-based firefighting (firefighting IoT) is still new to many countries. Currently, firefighting IoT mainly stays inside the enterprise, and no citywide system has been established. In this field, Chinese enterprises are at the forefront of the world. In 2006, the world's first provincial firefighting IoT system was established in Zhejiang Province. Since then, Chinese companies such as Tsinghua Tongfang and GST (Haiwan) have gradually matured in technology, market, and standard-setting capabilities. According to the incomplete statistics, as of the end of 2011, 28 provinces and municipalities of Mainland China have established remote monitoring systems for the IoT in firefighting protection.

6.2 Market analysis and social value

6.2.1 People's lives and property protection

China is a country with substantial fire losses. According to the statistics of the Fire Bureau of the Ministry of Public Security of China, in 2011, a total of 125,402 fire accidents were reported, 1106 people were killed, 572 were injured, and direct property losses were 1.88 billion yuan. Many of the fire losses were due to the poor quality of firefighting equipment and regulatory issues. Vigorously strengthening the construction of firefighting IoT, network-based fire control systems, and control devices are essential to reducing fire losses. With the rapid development of China's economy and society, various economic organizations and social units are constantly emerging. Specifically, more than 50 million organizations have been registered. Meanwhile, national fire supervision and the police force are only a little more than 10,000. Thus, China has to rely on technical means such as the IoT to improve supervision efficiency and minimize the damage caused by fire accidents.

The firefighting IoT platform can transmit the accurate fire alarm information received to the 119 fire command centers in real-time. This platform will fundamentally solve the situation of late reporting, concealment, and nonreporting of fire accidents, speed up the alarming speed and accurate judgment of the ignition points and reserve precious time for firefighting. At the same time, video surveillance can provide the best evacuation routes for the scene, reduce the actual loss of the fire, strengthen the management of the fire protection system, increase the rate of fire engines in the entire city, and improve the management functions and efficiency of the fire management centers. In addition to managing the equipment, under the authorization of the fire supervision and management department, it can also help manage the fire watch personnel on the networked units. The monitoring center can randomly check the duty situation of the local duty personnel and statistically calculate the duty and missing duty of fire prevention staff. By analyzing the failure data of each key unit, it is possible to calculate and estimate the manufacturer's product quality statistically, construction quality, and after-sales support capabilities, check the unit's fire management level, and supervise its improvement of fire management capabilities. With the implementation of the smart firefighting system, the annual inspection of the fire protection system can be more effectively implemented. An effective fire protection information management mechanism has been established, reducing the fire system failure rate, false alarm rate, 119 platform telephone alarm rate, and reducing the fire loss rate by 20%.

6.2.2 Water saving

According to the statistics, the total annual runoff of rivers in China has reached 26.38 billion cubic meters. However, due to its large population and extensive arable land, the per capita water volume is only 2700 cubic meters, ranking 88th globally. Water per square meter of arable land is only 2.65 cubic meters, about 3/4 of the world level. Among the 476 cities in China, 300 are short of water, about 15 million cubic meters daily.

Firefighting requires high water demand. According to the "Regulation for Fire Protection Design of High-Rise Civil Buildings," high-rise buildings must include the design of independent firefighting pools. Generally, high-rise buildings must provide 800–1200 cubic meters of firefighting pools. To ensure water quality and avoid damage to firefighting equipment, the water in the pool must be replaced regularly.

The problem of waste in the water is also severe. According to the statistics of the Water Saving Office of Hefei City, Anhui Province, the annual urban water consumption in Hefei is 290 million tons. Just because of the "run off-air, seepage, drip, leak" of the municipal fire hydrant and the municipal pipe network, 37 million tons per year of water are lost for nothing.

Water for firefighting has a bearing on people's lives and property safety and must be guaranteed. In addition, water is a crucial resource. Making full use of resources and minimizing water consumption while ensuring safety is a problem solved by the IoT of firefighting. The firefighting IoT can prevent water shortages from fire protection through technical standards, fire protection construction and acceptance standards, and research and development of water-saving fire-extinguishing equipment (Vajdová et al., 2019). Through the promotion of networked fire-extinguishing devices, the traditional practice of water-spraying fires can be gradually replaced.

6.2.3 Employment promotion

The construction of firefighting IoT is a comprehensive project, which requires many talents in various fields, such as sensors, control systems, networks, and software. In addition to research and development, there will be many employment opportunities in the production of fire equipment, networked monitoring operations, and value-added services. With the drive of "Industry-University-Research Collaboration," more than 700 universities and colleges have applied to launch the "Internet of Things Engineering" program. According to the statistics, the scale of China's Internet of Things industry in 2010 was close to 200 billion yuan, which has a huge driving effect on China's employment.

6.2.4 Intellectual property products innovation and promotion

In the "Twelfth Five-Year Plan for the Development of the Internet of Things," it was proposed that "the role of enterprises should be brought into place, and the current research results of universities and research institutes should be actively used and implemented in the field of information perception technology and information processing technology. The usage and implementation are to catch up with the advanced global level. to reach the advanced global level in the field of information transmission technology, to enhance information security guarantee capabilities, and strive to break through key core technologies as soon as possible to form a complete Internet of Things technology system and achieve industrialisation." In the field of firefighting, the gap between China and foreign countries is not significant. Chinese companies occupy the vast majority of the market, and the GST (Haiwan) company solely accounts for one-third of the national market. Chinese companies are fully capable of core technology research and development and product innovation.

"Firefighting IoT Platform" completely adopts China's intellectual property technology. Its core DCPL field control bus technology and network fire suppression technology have won Chinese national invention patents. Among them, the network fire extinguishing technology takes into account the shortage of water resources in China. It adopts the waterless fire extinguishing technology, which is the first in the world.

6.3 Overall architecture of system design

The firefighting IoT platform is an essential part of the comprehensive information portal of the city's IoT system. This platform can directly monitor the fire situation of various firefighting units in the city and monitor the fire situation of urban public transportation through wireless networks. The platform also has linkage capabilities: when a fire occurs, not only the fire department can be notified on time, but the fire can also be notified to the medical emergency, insurance, government emergency command centers, and other departments on time according to the plan, as shown in Fig. 6.1.

From an operational point of view, the city firefighting remote monitoring system is divided into three levels: the IoT awareness layer, network layer, and management service application layer, as shown in Fig. 6.2.

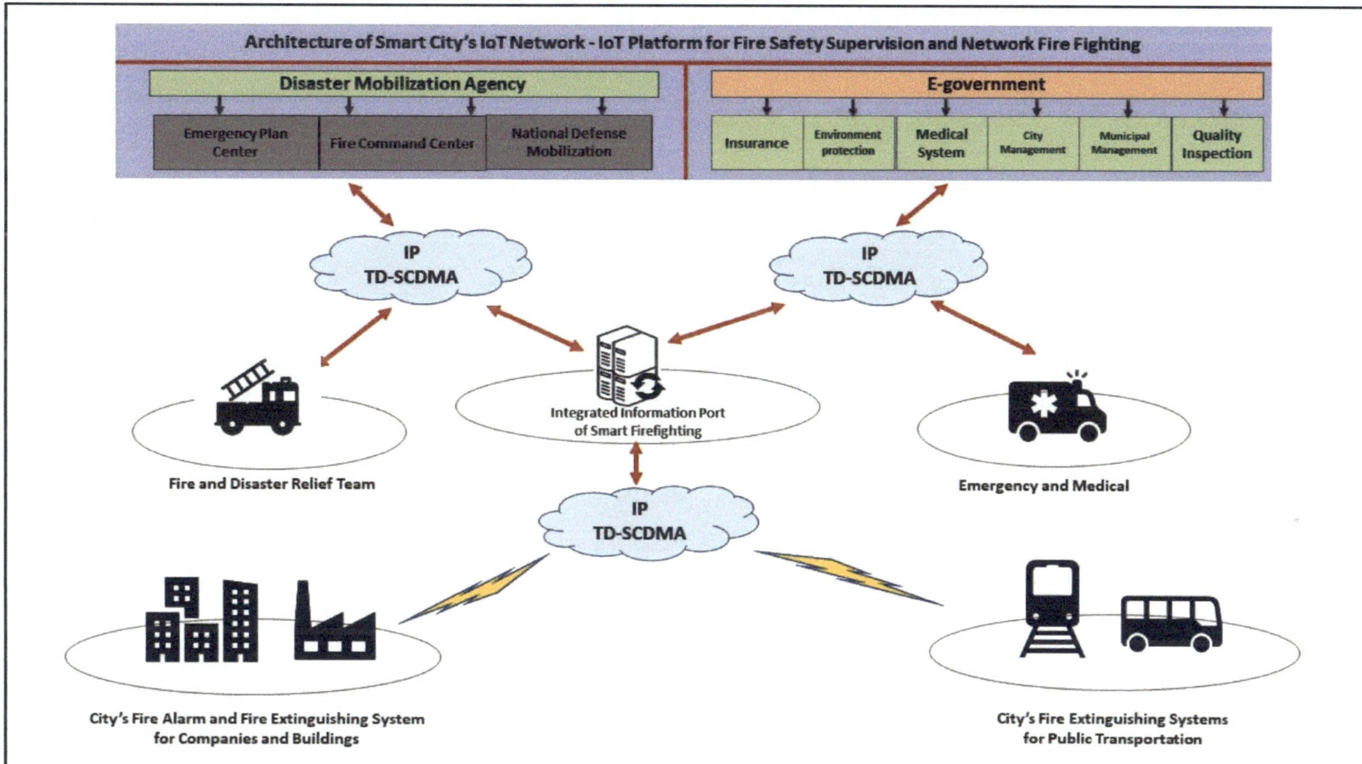

Fig. 6.1 Firefighting IoT platform.

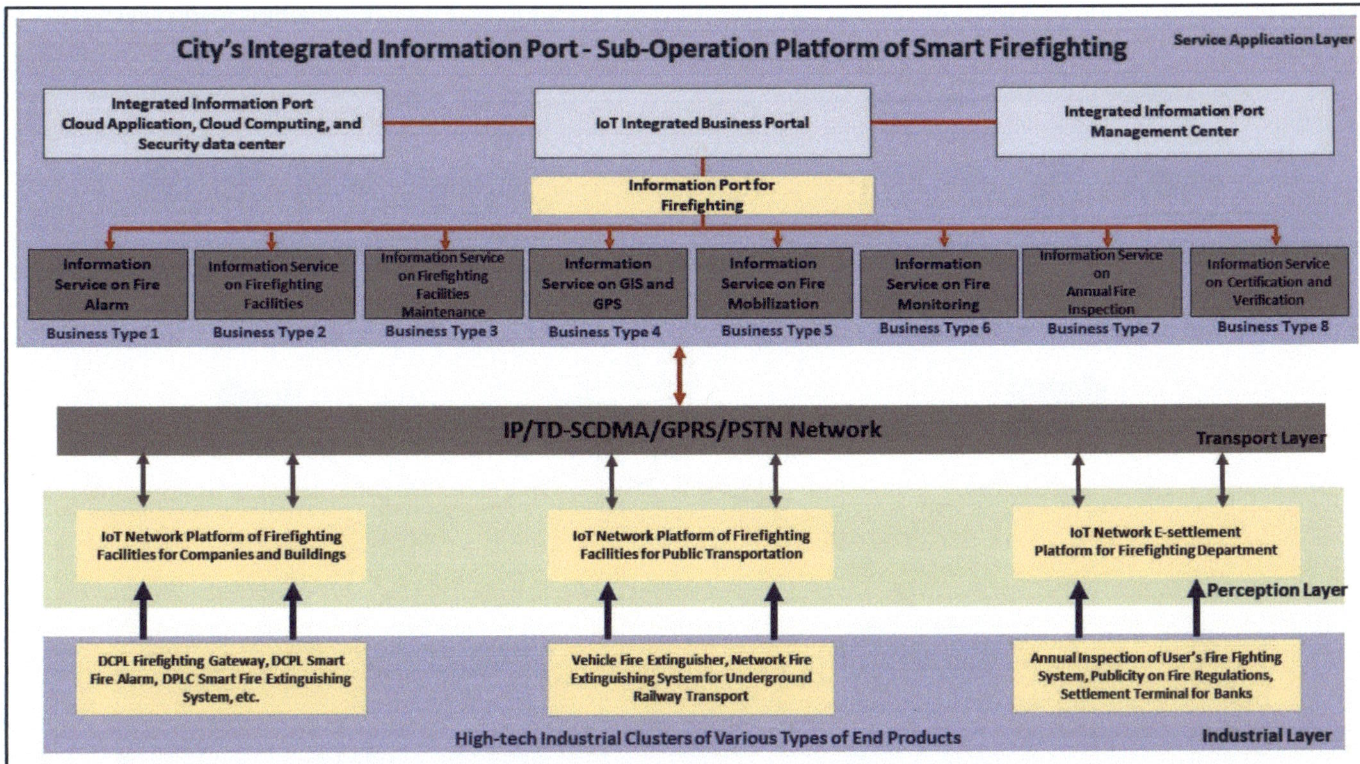

Fig. 6.2 Three levels of city firefighting remote monitoring system.

6.4 Smart city firefighting IoT solution

The primary role of the firefighting IoT platform is to monitor and extinguish fires through sensors and on-site control systems. The IoT concept mainly involves two parts:

(I) Fire extinguishing network system: The system is a complete network monitoring and fire extinguishing system, safe, reliable, and water-saving. The controller of the network fire extinguishing system can be directly connected to the city's IoT system. Thus, the fire monitoring system in the new buildings is recommended. Fig. 6.3 shows the structure of the fire extinguishing network system, and Fig. 6.4 shows the devices of the fire extinguishing network system.

(II) Firefighting IoT gateway: Most existing buildings nowadays are equipped with fire monitoring systems. To protect existing investments, the "IoT of Firefighting" supports the integration of existing fire monitoring systems. There is no unified standard for the equipment of mainstream manufacturers in the fire protection industry. The manufacturers themselves design the bus technology and control protocols. Therefore, a gateway device is required to convert the communication protocol of the existing fire monitoring systems. The fire IoT gateway is like an edge router on the Internet. It can make the decentralized and different standards of fire monitoring systems connected to a unified city firefighting IoT to achieve the purposes of both protecting existing investments and achieving networked monitoring.

6.5 Smart city firefighting IoT network layer

The network layer is the pipeline for data communication between the fire control system and the firefighting IoT monitoring center in the building. It can use multiple network environments such as wired or wireless and networks of multiple operators. Fig. 6.5 shows the operation of a city firefighting center.

The firefighting IoT platform requires the network layer to support a smart city. The network should be at least a standard of CDMA 1X or CDMA 2000 or WCDMA or GPRS or TD. Also, it needs to support PSTN and ADSL and the use of Internet exits on existing LANs to transmit data.

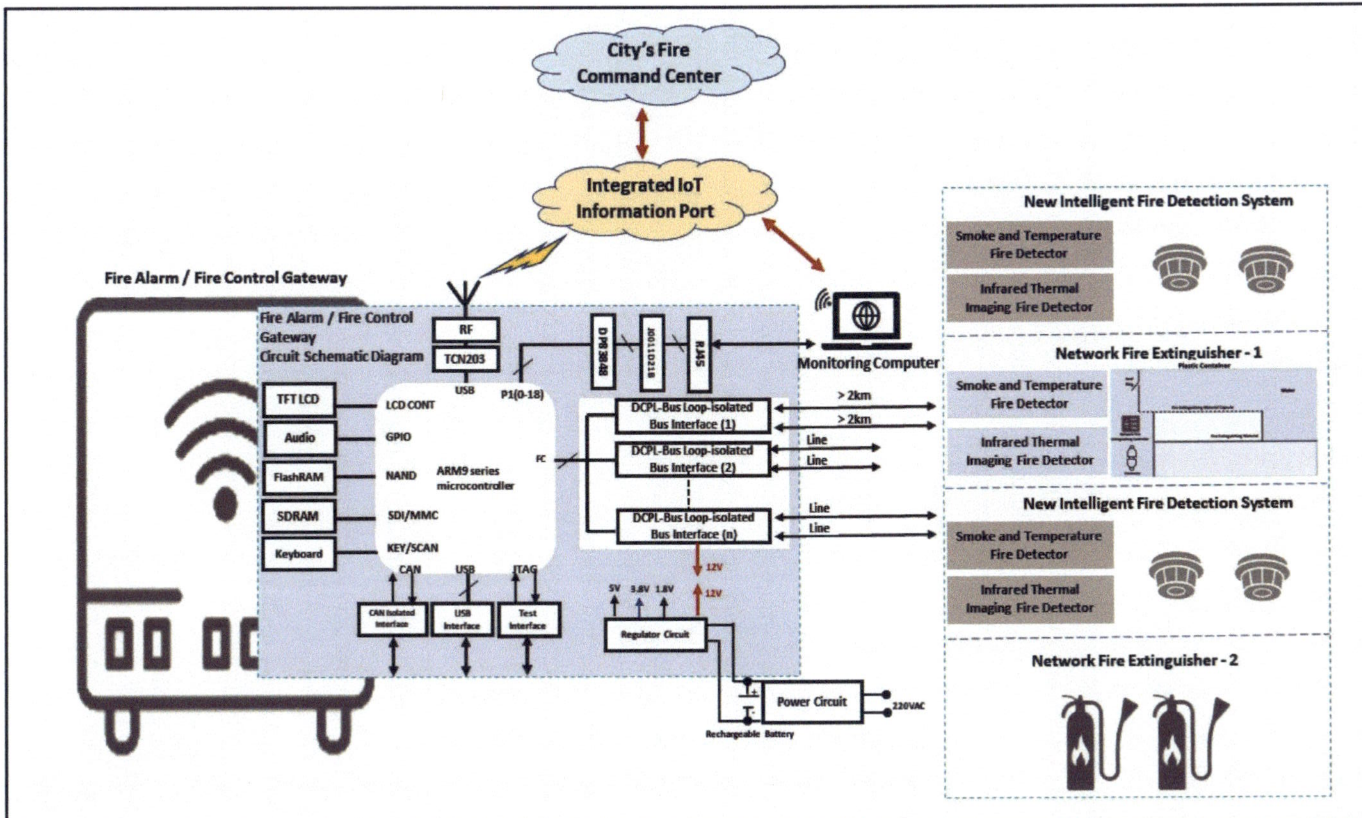

Fig. 6.3 Structure of the fire extinguishing network system.

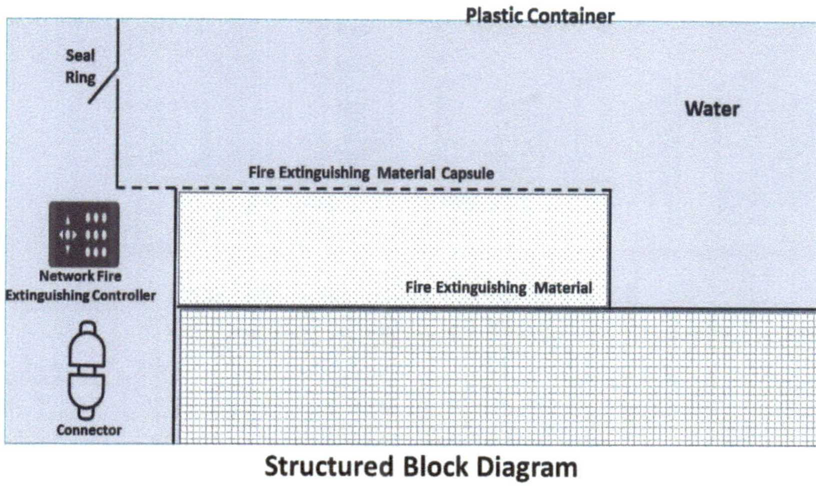

Fig. 6.4 Devices of the fire extinguishing network system.

6.6 Firefighting IoT service management application layer

The service management application layer is based on the city's public network. Amid the dynamic environments of various crises (Netten et al., 2016), this service management application layer relies on modern high-tech methods to remotely monitor the real-time operation of the automatic fire alarm system distributed throughout the city, to centrally monitor and manage the fire protection equipment of urban buildings, and to improve the operational efficiency and ability of the facility units of fire protection networked to withstand fires. Its main functions include receiving real-time alarm information and image information around the alarm point sent by the user-end monitoring device and analyzing and determining suspected and real fire alarms through the field's data, images, and voice information. Although the diverse building environment creates difficulty to fire prediction (Cheng et al., 2017), the monitoring center automatically distinguishes actual and faulty fire alarms according to the received information. After confirming the fire alarm information according to the suspected fire alarm and accurate fire alarm information, it processes the alarm information according to the

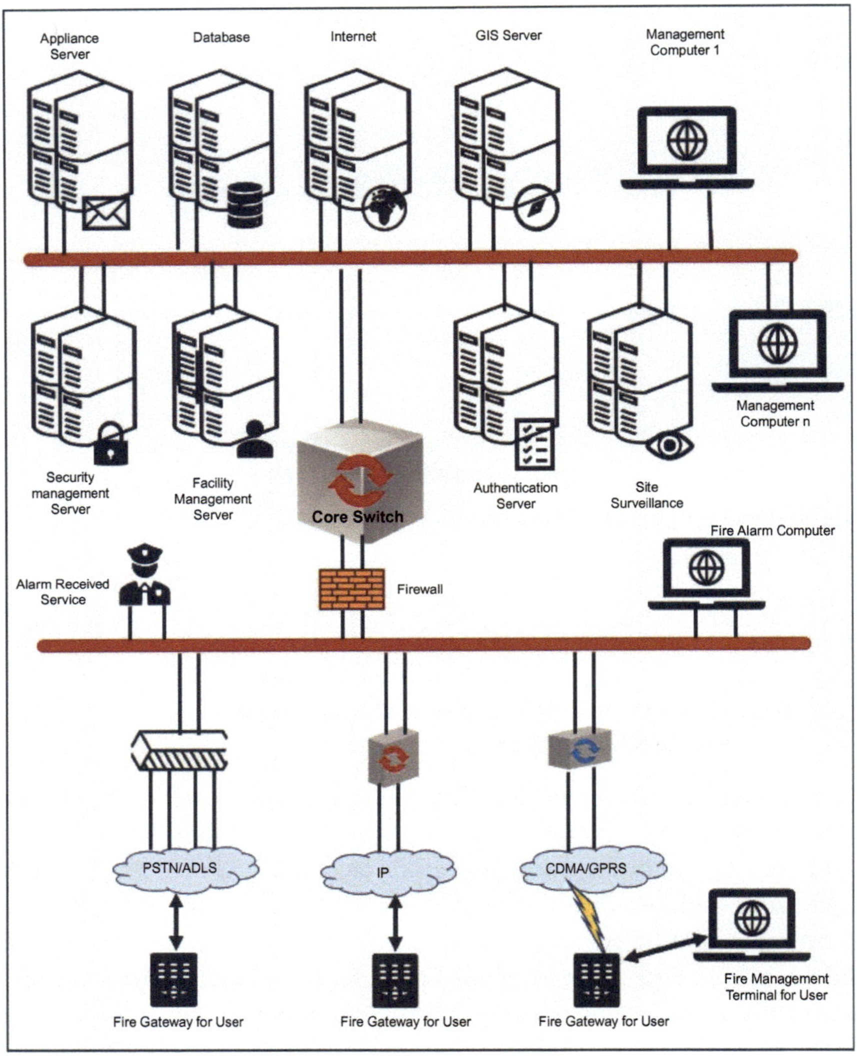

Fig. 6.5 Operation diagram of a city firefighting center.

corresponding policies. It will notify the firefighting and rescue department and other relevant departments as soon as possible for accurate fire alarm information. According to GB 50440-2007 "Technical Specifications for City Firefighting Remote Monitoring System," the service management application layer needs to meet the following requirements.

6.6.1 Functional requirements

After receiving fire alarm information from one of the networked users, the service management application layer will transmit the confirmed fire alarm information to the city fire command center or other receiving police centers, and it should be able to receive information on the operation status of fire protection facilities of the buildings from networked users. The management application layer should provide a query function to the City Fire Department to know about the fire alarm information of users, operation status information of the building's fire facilities, and information on fire safety management. The management application layer should also provide users with their fire alarm information, facilitate their query about the building's fire facilities operation status and fire safety management information. In addition, it should also be able to real-time update the operation status of fire protection facilities in buildings and fire safety management information sent by networked users. The most effective measures can be taken in the shortest time to reduce various losses.

6.6.2 Performance requirements

(1) Supporting multiple users: The monitoring center should receive and process fire alarm information for no less than three networked users simultaneously.
(2) Response should be timely: the response time from the acquisition of the fire alarm information from the user information transmission devices to the recipients and display of the monitoring center should not be greater than 20 s.
(3) Judgment must be quick: The time for the monitoring center to forward the confirmed fire alarm information to the urban fire communication command center or other receiving police centers should not be greater than 3 s.
(4) Constant inspection: The communication inspection cycle between the monitoring center and the user information transmission devices should not exceed 2 h. The inspection mode and time can be set dynamically.
(5) Backup: Records of the fire alarm information of the monitoring center and the operation status information of the building's fire protection facilities should be backed up, and its storage period should not be less than 1 year. When statistical processing is performed annually, it should be saved to storage media such as optical discs and magnetic tapes.

(6) Data should be completed: the storage period of the recording file should not be fewer than 6 months.
(7) Time should be consistent: the remote monitoring system should have unified clock management, and the cumulative error should not be more than 5 s.

In this chapter, the software of operation center management in the service management application layer should include the following functional modules:

(1) Video surveillance function

When a fire accident is reported, if there is a camera around the fire alarm site, the administrator should monitor the actual situation of the fire scene and determine whether it is a real fire alarm or a false fire alarm through the video on the scene. If it is a real fire alarm, the administrator will immediately notify the fire department and provide relevant data, such as the unit and location that sent the fire alarm, the situation of the building, the number of floors in the building, and the floor plan of the fire alarm location.

(2) Alarm module

When a fire alarm occurs, the alarm module can promptly notify the administrator of the monitoring center. The alarm information needs to be reflected on the work desk of the alarm administrator within 3 s. The system needs to inform the administrator of the fire occurrence unit and the exact location of the fire occurrence, such as displaying the name of the fire occurrence unit, the communication method, the size of the unit, the fire history of the unit, the floor plan of the site, the number and location of alarm probes, location of nearby cameras, etc. This module will also provide an accurate probability prediction on real fire for each fire alarm reported to the monitoring center according to specific rules. This probability can enable the administrator to have a rough idea to judge the incoming fire alarm. The algorithm of accuracy probability has the following basic principles: (1) The reliability of manual alarm information is high, and the probability of false alarm is small. (2) The fire emergency response measures of the building, such as lowering the elevator and opening the air outlet, are considered high credibility. (3) Simultaneous alarms of a series of adjacent probes have high reliability. If there is only one probe in a room, and this probe alarms, the reliability is considered relatively high. (4) Similar principles can be improved based on experience in practical applications to improve the accuracy of the complete system analysis. After receiving the alarm information, the administrator can learn about the fire through the camera or telephone. After confirming the fire, the administrator will notify the fire department immediately. If it is a false alarm, it is necessary to mark the alarm information as a false alarm and fill in the reason for the false alarm.

(3) Comprehensive information query

The comprehensive information query mainly conducts business inquiries on alarms, faults, user conditions, annual inspections, and payment conditions and supports the monitoring center's daily operations. The customer service staff and daily business processing staff of the monitoring center are the main users of this module.

(4) Report and data mining

The fire command centers have accumulated a large amount of historical fire data. Analyzing the accurate and useful information from the massive fire data to better upgrade the future fire command work is the main issue. Reporting and data mining modules can provide data analysis and processing functions and predict future development trends based on historical data. Making statistics and summary of fire alarm and fault information, and generate some detailed accident statistics record form, such as statistics record of high-incidence fire zones, hidden danger statistics record, fire management record of each key unit (for example, staff on duty, the monthly number of failures, number of false alarms, number of shutdowns, number of mains power outages, etc.), statistical reports of failure rates of automatic alarm systems per manufacturer, and after-sales service status reports per service provider, etc.

(5) Fault management system

The gateway of the networked units will report the monitored fault information in real-time. The fault information mainly includes the fire system equipment failure, system shutdown, etc. This module can classify and summarize all faults and prompt the administrator to eliminate the faults in time. If the networked unit encounters particular circumstances, such as decoration, it is necessary to temporarily turn off the fire IoT gateway. It must apply to the fire department and take the fire department's approval materials to the fire network monitoring center for registration. Otherwise, the system will treat it as an abnormal shutdown and process accordingly.

(6) Buildings and fire protection facilities management system

The buildings and fire protection facilities management system mainly focuses on the registration and management of conditions of the buildings connected to the network and the surrounding of fire protection facilities and dangerous goods (e.g., unit number, unit name, unit address, target objectives, adjacent units, phone numbers, contact persons, responsible fire stations and its driving routes, production, and storage of materials, types, and heights of buildings, key locations, fire hazards, geographic locations, fire plans, building plans, three-dimensional views, internal structure maps, etc.), for inquiry services to fire management departments and decision-makers.

(7) Portal management system for supervisors

Decision makers mainly use the portal management system for supervisors in fire management departments. Decision makers can log in to the platform through the authorization process anywhere and have real-time supervision on various information of the authorized jurisdictions (such as significant fire alarm events, historical alarm conditions, daily, monthly, and annual statistical reports, etc.). They can automatically receive various reports generated by the remote monitoring center to understand the overall situation of the jurisdiction fully.

(8) Annual fire inspection management.

Networking monitoring of key industries (such as Internet cafes, hospitals, etc.) is part of the annual fire inspections. Fire networked units need to present access proof to the networked system during annual inspections. The annual fire inspection management mainly assists operators in handling the related business during the annual inspections.

(9) Charge management

Internet users need to pay a certain amount of service fees each year. The main functions of charge management are to record the user's payment situation, print various reports, summarize financial status, and provide functions such as information queries.

6.7 Conclusion

The existing firefighting monitoring system can only provide monitoring functions and extinguish fires. Sprinkler systems also have the disadvantages of wasting water resources, being unable to be remotely controlled, and high construction and maintenance costs. Network fire extinguishing technology and firefighting IoT network system are considerable breakthroughs in this field. It has genuinely achieved waterless fire extinguishing and network remote control. As forward-looking, DCPL fieldbus and IoT data mining technologies will be necessary to further development in this field.

DCPL fieldbus technology: DCPL is a two-wire, active, direct current, analog signal fieldbus technology. It has the characteristics of low price and low power consumption. It is particularly suitable to be used as a fieldbus in fire protection, smart meters, and smart appliances.

IoT data mining technology: Through data mining, you can use statistical methods to analyze the causes of fires, the relationships between fires and regions, fires and fire equipment status, fires and weather, as well as fires and industries. After obtaining this information, it is possible to carry out fire prediction and early warning for the local area and eliminate fire hazards in advance.

In addition, like many other IoT applications in smart cities, firefighting IoT network systems has brought about a unified operating platform and multiprotocol integration technology. Adopting a unified platform can reduce the repeated development of the IoT network system and reduce the cost of implementing a firefighting IoT network.

References

Cheng, M. Y., Chiu, K. C., Hsieh, Y. M., Yang, I. T., Chou, J. S., & Wu, Y. W. (2017). BIM integrated smart monitoring technique for building fire prevention and disaster relief. *Automation in Construction*, *84*, 14–30.

Cicioğlu, M., & Calhan, A. (2020). Internet of things-based firefighters for disaster case management. *IEEE Sensors Journal*, *21*(1), 612–619.

Gaur, A., Singh, A., Kumar, A., Kulkarni, K. S., Lala, S., Kapoor, K., Srivastava, V., Kumar, A., & Mukhopadhyay, S. C. (2019). Fire sensing technologies: A review. *IEEE Sensors Journal*, *19*(9), 3191–3202.

Mahgoub, A., Tarrad, N., Elsherif, R., Ismail, L., & Al-Ali, A. (2020). Fire alarm system for smart cities using edge computing. In *2020 IEEE international conference on informatics, IoT, and enabling technologies (ICIoT)* (pp. 597–602). IEEE.

Netten, N., van den Braak, S., Choenni, S., & van Someren, M. (2016). A big data approach to support information distribution in crisis response. In *Proceedings of the 9th international conference on theory and practice of electronic governance* (pp. 266–275).

Ryu, C. S. (2015). Iot-based intelligent for fire emergency response systems. *International Journal of Smart Home*, *9*(3), 161–168.

Salhi, L., Silverston, T., Yamazaki, T., & Miyoshi, T. (2019). Early detection system for gas leakage and fire in smart home using machine learning. In *2019 IEEE international conference on consumer electronics (ICCE)* (pp. 1–6). IEEE.

Sharma, A., Singh, P. K., & Kumar, Y. (2020). An integrated fire detection system using IoT and image processing technique for smart cities. *Sustainable Cities and Society*, *61*, 102332.

Vajdová, I., Mikula, B., Jenčová, E., & Mako, S. (2019). The innovations improving fire services at airports. In *2019 new trends in aviation development (NTAD)* (pp. 198–203). IEEE.

Wu, X., Dunne, R., Zhang, Q., & Shi, W. (2017). Edge computing enabled smart firefighting: Opportunities and challenges. In *Proceedings of the fifth ACM/IEEE workshop on hot topics in web systems and technologies* (pp. 1–6).

Zhang, G., Yan, S., Zhu, G., Feng, P., & Jia, B. (2020). Smart firefighting construction in China: Status, problems, and reflections. *Fire and Materials*, *44*(4), 479–486.

CHAPTER SEVEN

Smart parking using mobile and IoT

7.1 Introduction

Parking coordinates transportation and land use in urban areas as one of the most critical assets to bring revenues to cities. Averagely, in big cities such as New York, San Francisco, Los Angeles, and Melbourne, parking coverage takes more than 31% of land use (Lin et al., 2017). In addition, due to the insufficient supply of parking slot, the drivers may spend a lot of time finding, which leads to traffic congestion and emissions (Xiao et al., 2018). McCoy (2017) stated that drivers in New York, on average, spend 107 h a year finding parking slots that estimate as $2243 in wasted time, fuel, and emission.

Hong Kong also faces parking problems, especially in built-up areas, due to limited land space and many cars. It is common to see that the roadside space is occupied by the logistics trade, such as loading-and-unloading activities, picking-up and dropping-off services, and various obstacles in Hong Kong (Cao et al., 2017; Ma et al., 2018). At the same time, the Hong Kong Government has been promoting green transportation and electric vehicles as solutions to reduce greenhouse gases emissions generated by motorized vehicles. However, due to the low transparency of public parking service information, it is difficult for drivers to find the right parking spot at the same time. The four "Right" requirements are the right location with the right services, such as a parking space with a charging facility for electric vehicles, is the four "Right" requirements.

Recently, Wireless Sensor and Actuator Network (WSAN), Radio Frequency Identification (RFID), Near Field Communication (NFC), Internet of Things (IoT), and cloud services have emerged as the new technologies. It can be applied in transportation to improve public parking services and reduce environmental pollution (Sabbea et al., 2018; Tsiropoulou et al., 2017). With the IoT, drivers can easily target and tackle a vacant parking

Collective Intelligence for Smart Cities
https://doi.org/10.1016/B978-0-12-820139-8.00003-6

spot for instant monitoring. In addition, this can reduce the chances of errors due to misinformation from individuals (Ni et al., 2018; Paidi et al., 2018). Due to the benefits offered from this technology, the concept of smart parking has been widely adopted in various countries such as the United States, China, and Japan (Chester et al., 2015; Grossman & Helpman, 2005). In considering the land limitation and costs in handling the traffic and transportation problems in Hong Kong, extending the smart parking concept in Hong Kong is crucial to provide efficient parking services cost-effectively.

Hence, an IoT-based advanced automobile parking navigation platform (IAPNP) is designed in this chapter, which integrates RFID, WSAN, NFC, IoT, and cloud computing. It aims to provide the platform with a high-accessible visual status of parking spaces and facilitate the decision making and service management for a service provider. Moreover, the proposed platform also provides the drivers with real-time enquiries on car park information to enjoy the high quality of parking environment. The rest of the chapter is organized as follows: The architecture of the proposed platform is described in Section 7.2. The case study is conducted in Section 7.3. Section 7.4 is the conclusion.

7.2 Architecture of the IAPNP

In this section, the IoT-based automobile parking navigation platform (IAPNP) is mainly composed of IoT infrastructure and software, middleware, hardware, and mobile app based on it.

7.2.1 IoT infrastructure and functions

Four layers of IoT structure are deployed in designing and implementing the proposed platform by integrating advanced, emerging, and innovative technologies, as shown in Fig. 7.1. Such technologies include RFID, WSAN, NFC, IoT, cloud computing, and so on. The sensor layer comprises equipment and technologies, such as sensors, actuators, radio frequency identification (RFID), and near-field communication (NFC). Communication technologies are relevant to the network layer, e.g., Wi-Fi and Long-Term Evolution (LTE). Critical technologies covered by the platform layer include cloud computing, virtualization, and big data. Lastly, the application layer includes a real-time search of parking spaces, reservation, car park navigation, parking space guidance, indoor vehicle location, parking space resource forecast, and planning functions. Under the IoT environment,

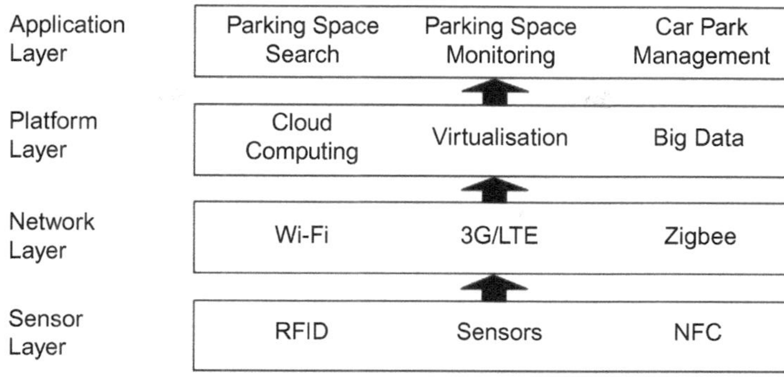

Fig. 7.1 IoT architecture.

the platform can share the collected data more effectively and efficiently to facilitate further data analytics and decision making. All of these are found in the field of intelligent transportation.

7.2.2 Core components and their functions

The proposed IAPNP consists of four major components: (i) WSAN for Smart Parking Management System, (ii) WSAN Middleware, (iii) Advanced Automobile Parking Navigation System, and (iv) NFC-enabled Customer Relationship Management Application. Fig. 7.2 shows the overall structure of the IAPNP. The proposed IAPNP provides convenience for both drivers and service providers. This system can capture data in real-time for the service providers for monitoring the car park environment and management. Through the performance analysis, appropriate planning and forecasting can be provided for the service providers to improve the service quality. For the drivers, this platform provides them with a full range of services, which

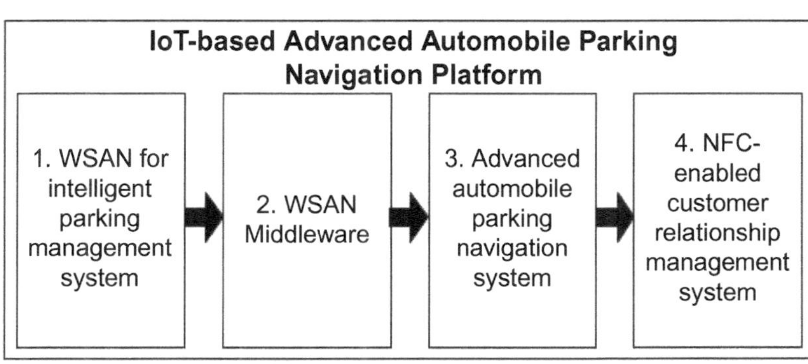

Fig. 7.2 Overall structure of the IAPNP.

cover online parking reservation enquiries, navigation, mobile customer management, a real-time search of parking spaces, electric car parking space reservation, indoor car positioning.

7.2.3 Main features and functions

7.2.3.1 WSAN for smart parking management system

The first system in the proposed IAPNP is WSAN for Smart Parking Management System, the hardware development stage. Four major components are involved in this system: Wireless Parking-vehicle Detection Node, LED indicators, Wireless EHS Sensor and Actuator Node, and WSAN Gateway with Embedded Operating System. The structure of WSAN for the Smart Parking Management System is shown in Fig. 7.3.

Two nodes, i.e., wireless parking-vehicle detection node and wireless EHS sensor and actuator node, are installed above each car parking space and around the car park, respectively, to collect real-time data. Such data are then transmitted to the WSAN gateway with the embedded operating system through a wireless network such as Wi-Fi. Through the Message Queuing by telemetry Transport (MQTT), the collected data can be handled, converted, transmitted to the MQ Server and subsequently the WSAN

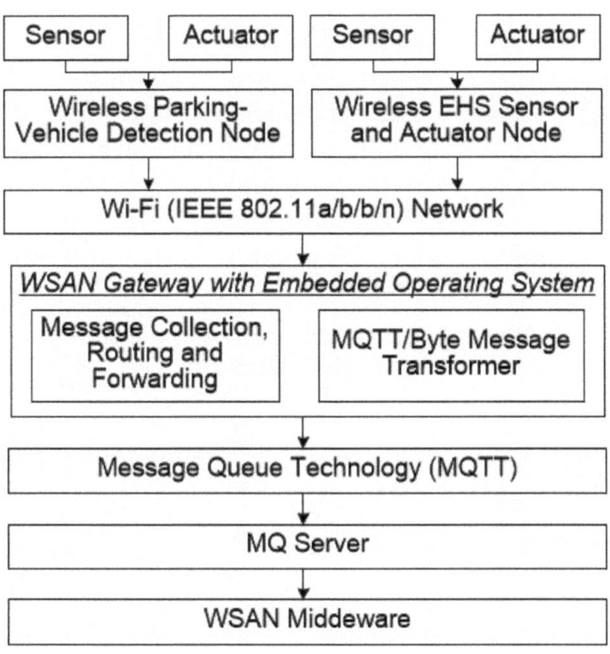

Fig. 7.3 Structure of WSAN for smart parking management system.

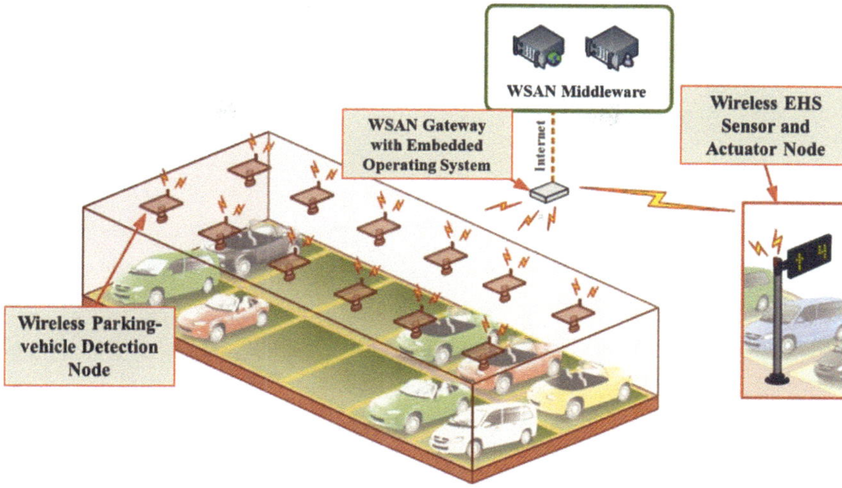

Fig. 7.4 A schematic diagram showing WSAN for smart parking management system.

middleware. Based on the information received from the sensor and actuator, the WSAN middleware will issue commands and generate the corresponding commands to the sensor and actuator to guide drivers to search out available parking spaces. Apart from the above function, this system also provides 24-h monitoring for automatically adjusting the car parking environment such as brightness, temperature, air quality.

In the wireless parking-vehicle detection node, the dual infrared distance detector and connected LED indicator lights are equipped to detect the occupation status of the parking space and show the visual signals, i.e., "Available", "Occupied" and "Reserved," for informing drivers of the occupation status, respectively (Fig. 7.4). Moreover, in the wireless EHS sensor and actuator node, four primary environmental sensors, i.e., brightness sensor, gas sensor, temperature, and humidity sensor, are equipped to achieve the functions of real-time environmental monitoring in the car park. By automatically turning to the light or turning on/off the temperature control device, custom conditions can reduce energy consumption and protect the environment. Figs. 7.5 and 7.6 show the structure of the two nodes.

7.2.3.2 WSAN-based middleware

With the data collected stored by the MQ Server in the WSAN for Smart Parking Management System, the WSAN-based middleware will process such data into processed commands through collecting, filtering, storing,

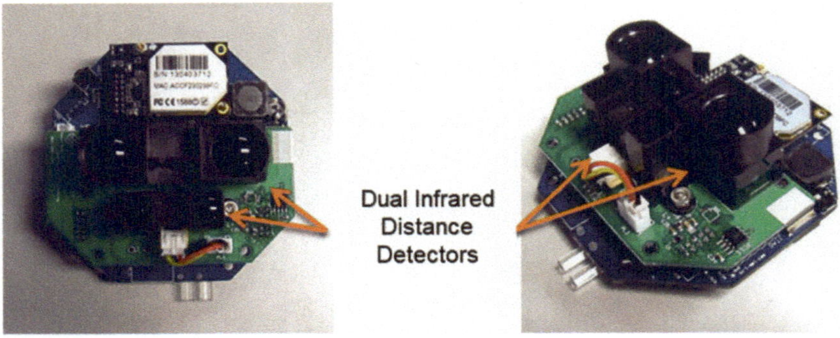

Fig. 7.5 Dual infrared distance detectors in the wireless parking vehicle detection node.

Fig. 7.6 Four types of sensors in the wireless EHS sensor and actuator node.

and aggregating. Fig. 7.7 shows the structure of WSAN middleware. After that, processed commands will be treated to issue messages to all parts of the system. Such as the wireless parking-vehicle detection node located above the parking space, the wireless EHS sensor and actuator node, and Advanced Automobile Parking Navigation System to display real-time information. Moreover, the user interface of WSAN middleware provides the

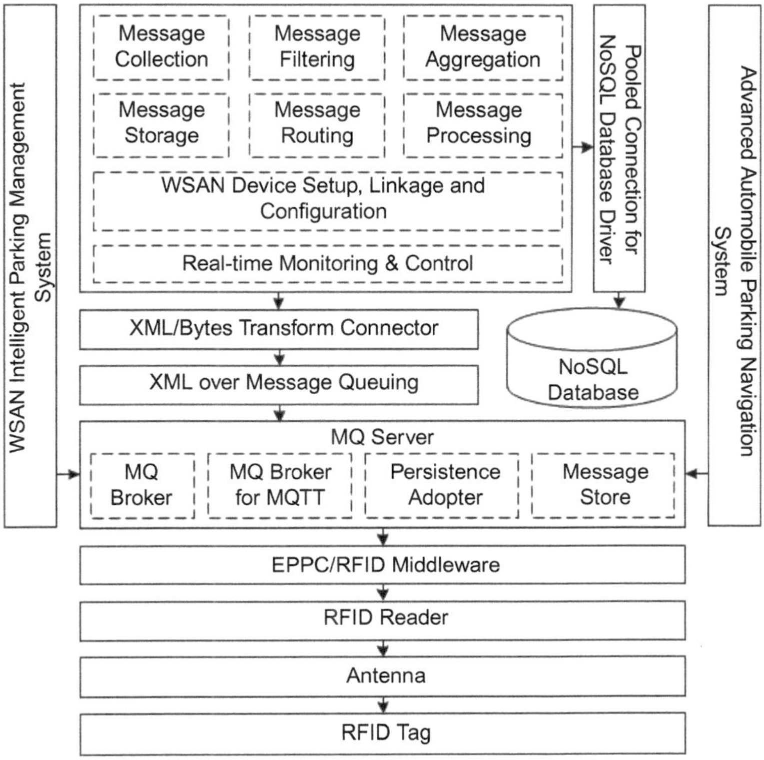

Fig. 7.7 WSAN middleware.

convenience for users to visualize and manage the components involved in the Advanced Automobile Parking Navigation System. Service providers can remotely control the status of detector nodes or the actuator, as shown in Fig. 7.8. As the adapter between software and hardware, the user interface of WSAN middleware can connect to different embedded devices and systems. Providing these functions improves the flexibility, availability, and scalability of the proposed IAPNP.

7.2.3.3 Advanced automobile parking navigation system

The Advanced Automobile Parking Navigation System is the core software development part in the proposed IAPNP which consists of two subsystems: (i) Car park operating system (CPOS) and (ii) Car park management system (CPMS). Two main functions, i.e., car park management and monitoring and quality environmental monitoring, can be performed in this system, as shown in Fig. 7.9. With the real-time information, the services providers

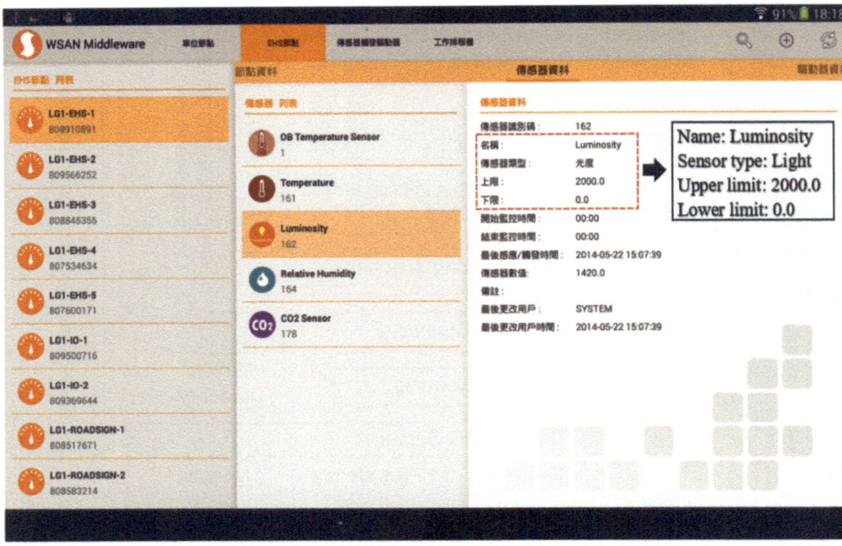

Fig. 7.8 User interface of WSAN middleware.

Fig. 7.9 User interface for car park management and real-time monitoring.

can easily display the required information related to the parking spaces, the environment, or the security. Apart from real-time monitoring, the environmental data can be recorded and analyzed through the system to facilitate service providers' decision-making.

The CPOS will be installed in the shroff office of the car park to bring convenience to the service providers in managing parking spaces. Through the CPOS, the staff can automatically collect parking fees by browsing information about individual transactions. This can speed up the payments service in the entire car park through the application of RFID technology. On the other hand, in the CPMS, all car park distribution and location information can be displayed. Service providers and the staff can use the data collected by the two nodes to generate the data analysis reports for a car park. This collected data assists the management in reviewing the operation circumstances of each car park and thereby increasing revenue through planning marketing strategies. Meanwhile, the database of this platform is also conducive to customer relationship management. Service providers can increase customer loyalty by launching different reward programs and building up long-term relationships with customers.

7.2.3.4 NFC-enabled customer relationship management mobile app

To maintain a good relationship with customers, the drivers can download the mobile app, namely NFC-enabled Sino Parking Mobile App, on their smartphone. Fig. 7.10 shows the structure of the NFC-enabled Customer Relationship Management Mobile App. This app integrates cloud technology, Advanced Automobile Parking Navigation System, and Google Map to provide a range of services for drivers. Drivers can access the car parking information, including the real-time search of available parking spaces, electric car parking space reservation, street map navigation, the electronic payment with NFC, and enquiries about customers' loyalty program information anywhere at any time.

7.3 Case study

A pilot test was conducted in a car park operated by Sino Parking Services Limited to validate the feasibility of the proposed IAPNP. Owning more than 120 car parks in Hong Kong, the case company aims to provide user-friendly, cashless payment methods and automated driver access systems. The pilot test provides important practical data for the future development of IAPNP and its large-scale promotion and adoption in Hong Kong.

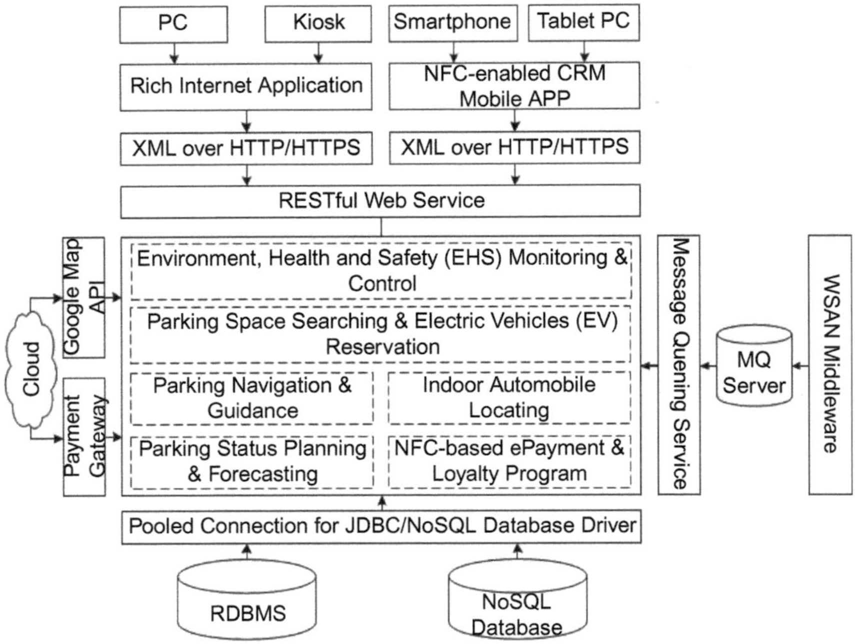

Fig. 7.10 NFC-enabled customer relationship management mobile app.

The car park chosen for the pilot was the car park at The Coronation. The car park is located at 1 Yau Cheung Road in southwest Kowloon. It provides various parking spaces, including residents' parking spaces, visitors' parking spaces, commercial vehicle parking spaces, motorcycle parking spaces, and lorry loading/unloading bays. Advanced vehicle parking navigation platform was successfully installed on the car park at The Coronation, and the 60 hourly parking spaces were also fitted with intelligent hardware for testing.

7.3.1 Implementation of the IAPNP

The proposed IAPNP and the corresponding software system, middleware, intelligent hardware equipment, and mobile app are fully installed to the case company for (i) delivering a range of quality parking services and (ii) providing a new experience for motorists parking lot users. Meanwhile, the proposed IAPNP significantly improved information transparency about the car park, providing the parking service supplier with advanced and visualization tools for monitoring parking spaces and decision-making. The following describes the actual application of this platform.

7.3.2 A new experience of quality parking services

The proposed IAPNP provides motorists with a new one-stop service experience for quality parking from the driver's perspective. They could use the Sino Parking app installed on their smartphones to search for a suitable car park nearby or adjacent to the destination. Drivers can immediately inquire about information because parking information for all car parks could be retrieved through the smartphone app (Fig. 7.11). This information includes the number of available parking spaces in the current car park, the hourly fees at car parks, complete address, photos of car park entrances, nearby facilities and other information, which are all shown for drivers to decide which car park to go to, as shown in Fig. 7.12. After a car park is

Fig. 7.11 User interface of the mobile app to show the car park.

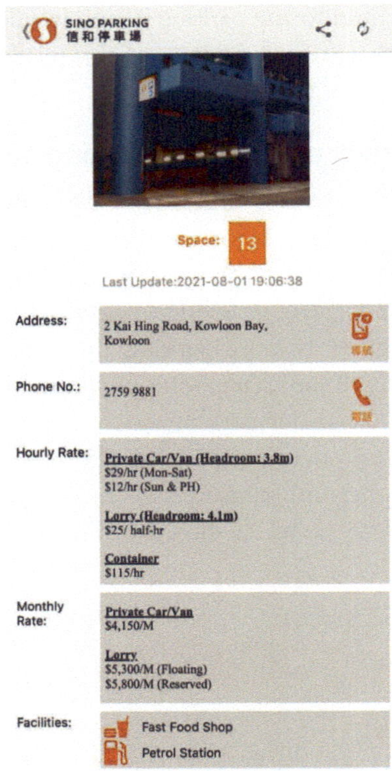

Fig. 7.12 Parking information shown in the mobile app.

selected, the smartphone app will even navigate the driver to the desired car park.

Upon arriving at the car park, the driver could use an NFC-enabled smartphone on the sensor at the entrance, as shown in Fig. 7.13. There is a display board in the car park to show the number of parking spaces in different sections. With such information, drivers can easily find a suitable parking space by following the direction shown on the display board, as shown in Fig. 7.14. In addition, an independent indicator light with different colors is installed in each parking space to display its availability. The red light indicates that the parking space was used, while the green light indicates vacant parking. In addition, through the mobile app, the driver could record the position of space used, making it convenient for getting to the car later. By simply following the route shown on the display board and looking at the status of indicator lights, the drivers will not miss any vacant parking space in the car park. Hence, they can enjoy the new experience of quality parking services.

Fig. 7.13 NFC technology at the entrance of the car park.

Fig. 7.14 Display board to show the number of vacant parking spaces.

7.3.3 Electric parking/enquiry and reservation services for parking spaces for people with disabilities

Apart from the general parking services, the drivers could use the mobile app's information to search the spaces for electric vehicle charging and people with disabilities. By calling the car park, the staff at the parking service provider will reserve the available space for the drivers. Once the parking space is reserved, the indicator lights will change from green to blue to provide convenience for drivers with special needs.

7.3.4 Quality environmental monitoring and management

To allow drivers to enjoy a quality environment and service at the car park, the intelligent sensors of the proposed IAPNP would not only update the parking status in real-time. Still, they would continuously monitor the environmental data, including temperature, humidity, air quality in the car park to facilitate enquiries by the management staff in the case company. Fig. 7.15 shows the user interface for monitoring the environment data. Once the ambient indices exceed the preset parameters, the system would send notifications in real-time to notify the management staff of the parking service provider to act accordingly. Moreover, the proposed IAPNP was also linked to devices that controlled the ambiance inside the car park, such as ventilation and lighting. Therefore, the adjustment could be automatically made

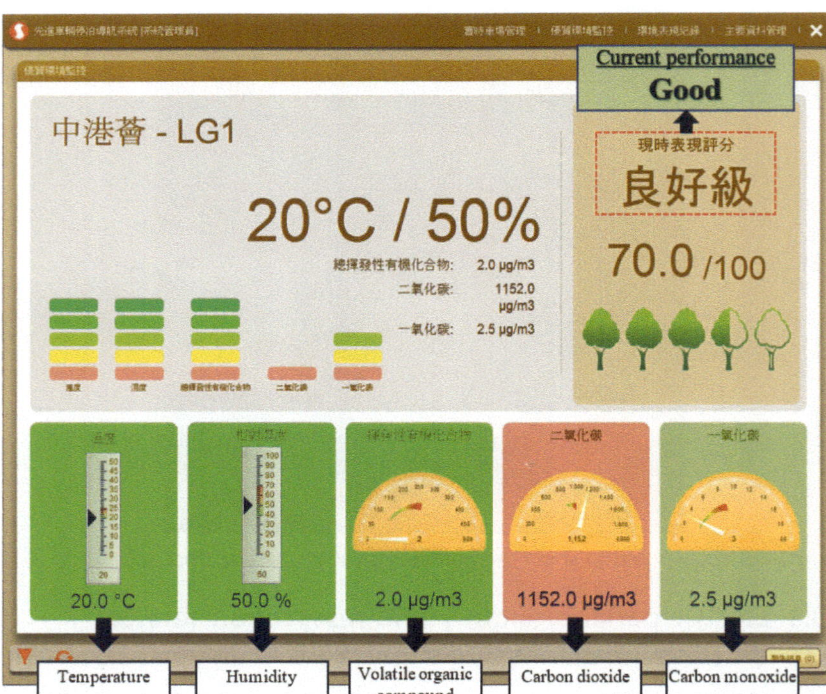

Fig. 7.15 User interface to monitor the environmental data.

based on the instruction provided by the management staff to provide a quality car park environment for users and improve service satisfaction.

7.3.5 Automated green energy-conserving and security management

To protect the environment and reduce greenhouse gas emissions, green energy conserving and security management were also deployed in the IAPNP. When usage is low or sufficient outdoor lighting in the car park, some of the lightings inside the car park will automatically turn off to save energy. Meanwhile, in providing convenience, lighting in a specific section of the car park would automatically turn on when someone triggered the sensors in that section. In addition, within a preset period (the evening or early morning before sunrise), when the sensor detected someone entering the car park, the system would alert the management staff to pay extra attention to the surveillance camera installed in the section concerned.

7.3.6 Real-time system analysis report

With the support of the IAPNP, a car park can function round the clock, 24 h a day. Through a dedicated network, the management staff can monitor various car parks from a "24-h central control center." The proposed IAPNP improves service performance and improves customer relationship management, human resources management, and quality management in the car park in terms of customer relationships. The management staff can obtain real-time information about parking and income profiles of each car park through this platform. As shown in Fig. 7.16, they can also obtain different usage reports on the car parks, which greatly assist the management in reviewing the operation of each car park and in subsequent planning of marketing strategies to increase revenue. The corresponding reward program could be provided for different customers to increase customer loyalty based on the analyzed results.

In addition, the system strengthens the efficiency with which the parking service provider operates. Based on the information and knowledge obtained from the IAPNP, they can instantly notify the number of parking spaces available in the car park and allocate the proper human resources to serve the customers. Furthermore, this system can control the ambient air quality and record data through the ambiance and security modules. If the air quality is unsatisfactory, property management personnel can

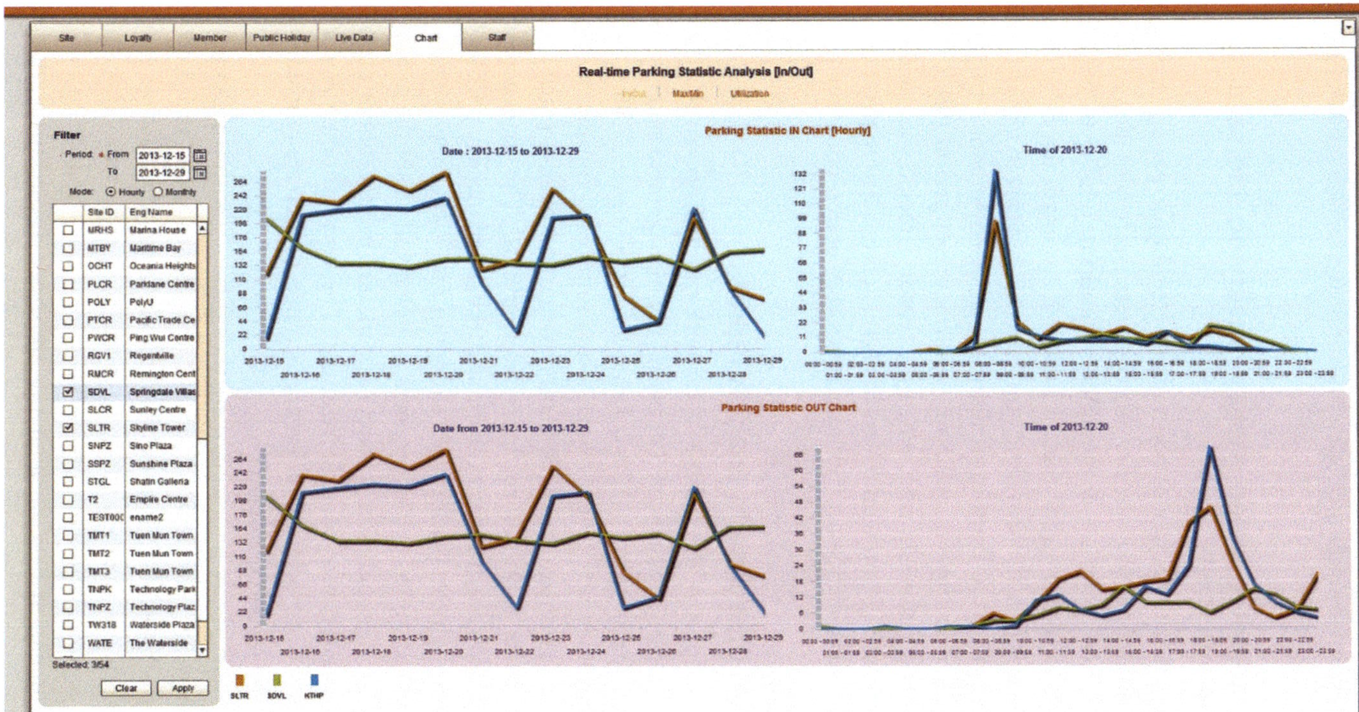

Fig. 7.16 User interface for data analysis.

immediately obtain such information and take appropriate measures to achieve continuous improvement.

7.4 Conclusion

Due to road traffic congestion and limited land use in Hong Kong, Hong Kong is a persistent problem that leads to unnecessary blocking and safety issues for all road users. Aligning with the national and Hong Kong's Smart City development, smart mobility with advanced technologies is a significant concern. The IoT-based advanced automobile parking navigation platform (IAPNP) is proposed to achieve highly efficient road networks. IAPNP consists of four main parts: the Sino Parking app, the advanced automobile parking navigation system, WSAN middle, and smart parking management systems. Users, i.e., drivers, can use smartphones to download the mobile app to perform a series of value-added customer services, such as searching for car parks nearby or in a specified area, getting real-time information about car parks, receiving navigation to the specified car park, recording and viewing parking spaces. The advanced automobile parking navigation system is connected to an NFC-enabled mobile app and WSAN middleware. The core component of the platform can achieve real-time search for parking spaces, advance scheduling of navigation and guidance, environmental monitoring and control, indoor vehicle positioning, among other functions. WSAN middleware serves as the bridge that connects the hardware and software of the system: while collecting and transmitting data, it also preprocesses the data by filtering, storing, collecting, aggregating and replying, etc., thereby reducing network traffic and the workload of the hardware and software. Within the WSAN for Smart Parking Management System, the wireless parking sensor is connected to the reactor. It can detect the parking status in real-time and display the statuses in different color lights, significantly improving the accessibility and transparency of information about parking spaces in public car parks. The wireless environment sensor is connected to the reactor. It transmits wirelessly the information collected about the environment to the operating system, which corresponds with effecting adjustments to the environments according to the information collected. The proposed IAPNP facilitates the decision making and service management for car park service providers through a better understanding of the collected data.

References

Cao, K. H., Cheng, Y. S., & Woo, C. K. (2017). Price-management of traffic congestion: Hong Kong's Lion Rock Tunnel. *Case Studies on Transport Policy*, *5*(4), 699–706.

Chester, M., Fraser, A., Matute, J., Flower, C., & Pendyala, R. (2015). Parking infrastructure: A constraint on or opportunity for urban redevelopment? A study of Los Angeles County parking supply and growth. *Journal of the American Planning Association*, *81*(4), 268–286.

Grossman, G. M., & Helpman, E. (2005). Outsourcing in a global economy. *The Review of Economic Studies*, *72*(1), 135–159.

Lin, T., Rivano, H., & Le Mouël, F. (2017). A survey of smart parking solutions. *IEEE Transactions on Intelligent Transportation Systems*, *18*(12), 3229–3253.

Ma, R., Lam, P. T., & Leung, C. K. (2018). Potential pitfalls of smart city development: A study on parking mobile applications (apps) in Hong Kong. *Telematics and Informatics*, *35*(6), 1580–1592.

McCoy, K. (2017). *Drivers spend an average of 17 hours a year searching for parking spots*. Available from https://www.usatoday.com/story/money/2017/07/12/parking-pain-causes-financial-and-personal-strain/467637001/.

Ni, J., Zhang, K., Yu, Y., Lin, X., & Shen, X. (2018). Privacy-preserving smart parking navigation supporting efficient driving guidance retrieval. *IEEE Transactions on Vehicular Technology*, *67*(7), 6504–6517.

Paidi, V., Fleyeh, H., Håkansson, J., & Nyberg, R. G. (2018). Smart parking sensors, technologies and applications for open parking lots: A review. *IET Intelligent Transport Systems*, *12*(8), 735–741.

Sabbea, M. O. B., Irfan, M., ALtamimi, S. K., Saeed, S. M., Almawgani, A. H. M., & Alghamdi, H. (2018). Design and development of a smart parking system. *Journal of Automation and Control Engineering*, *6*(2).

Tsiropoulou, E. E., Baras, J. S., Papavassiliou, S., & Sinha, S. (2017). RFID-based smart parking management system. *Cyber-Physical Systems*, *3*(1–4), 22–41.

Xiao, H., Xu, M., & Gao, Z. (2018). Shared parking problem: A novel truthful double auction mechanism approach. *Transportation Research Part B: Methodological*, *109*, 40–69.

CHAPTER EIGHT

Crane selection for project cargo

8.1 Introduction

The role of the logistics industry has been raised as it facilitates the growing international trade (Delfmann et al., 2018; Rai et al., 2012). Global outsourcing enables a rising economy of scale, hence the need for heavy and oversized shipments for projects (Grossman & Helpman, 2005; Melnyk & Malaksiano, 2020). Construction and many industries, such as automobiles, aviation, manufacturing, entertainment, and even performing art, need to transport oversized objects. In 2014, about 24.5% of the total international seaborne trade was contributed by project cargo (UNCTAD, 2015). Unlike general cargo ranging from 20 fleets to 40 fleets, project cargo operation is more complex and customized as the high fluctuation nature of the project. Project cargo is heavy and massive, which common equipment such as open-top containers and flat racks cannot handle (Lindstad et al., 2016). Special equipment, i.e., mobile cranes, is needed for loading and unloading, packing and unpacking, and installation in the destination (Şakar et al., 2018). Currently, there are two problems in the mobile crane selection process.

Firstly, due to the special handling requirements for various project cargo orders, it is time-consuming for workers to search for such information manually. In addition, considering workers with different experiences in handling project cargo orders may affect the quality of the mobile crane selection. The improper crane used to handle the oversize cargo may lead to product damage and even loss to companies. Recently, the Internet of Things (IoT) has drawn attention to achieve smart construction and logistics to improve the effectiveness, security, and safety issues of crane applications (Zhong et al., 2014). For example, a wireless sensor can continuously measure and transmit load orientation from the crane to the smart devices for data processing (Fang et al., 2016). Therefore, a smart crane selection system (SASS) is developed to streamline the crane selection processes in logistics companies. The data management tool gathers relevant data from various

Collective Intelligence for Smart Cities
https://doi.org/10.1016/B978-0-12-820139-8.00012-7

parties and then transfers it to a cloud-based centralized database. After that, the artificial intelligence technique, i.e., analytic hierarchy process (AHP), is employed to identify critical factors affecting the crane selection and provide decision support for workers. By conducting the pilot study in the case company, the SASS helps improve the operation performance in handling project cargo orders and hence increases service satisfaction. The rest of the chapter is organized as follows: Section 8.2 presents the research methodology of the proposed SASS. A case study is conducted in Section 8.3, while the results and discussion are presented in Section 8.4. The conclusions are drawn in Section 8.5.

8.2 Research methodology

A smart crane selection system (SCSS) is proposed to facilitate the mobile crane selection for project cargo. The system consists of Module 1: Data Collection and Preparation, Module 2: Analytic Hierarchy Process, and Module 3: Evaluation. Fig. 8.1 shows the architecture of the proposed SCSS. First, all relevant data from various parties are collected and stored in the centralized data warehouse in the cloud. Module 2 is responsible for analyzing the importance of crucial criteria in the mobile crane selection using AHP. To evaluate the appropriateness of the analyzed results, a questionnaire is designed in module 3 for workers to collect their feedback after implementing the suggestions from module 2. Details of each module are presented in the following sections.

8.2.1 Module 1: Data collection and preparation

In module 1, capability, economic, safety, and flexibility data are collected from logistics companies, crane manufacturers, and crane renting companies. Table 8.1 shows an example of data collected for the mobile crane selection.

Considering that these kinds of data are collected from various parties, there is variability in the data sources and data heterogeneity. Therefore, data preprocessing is an essential step for data preparation. The methods of data cleaning, preprocessing, and integration can enhance the quality of collected data. For example, it can remove unclear, duplicated and incomplete data. Also, it can combine data from multiple heterogeneous data sources into coherent data and generate a unified view of data. After the data preprocessing step, valuable data are stored and transferred to the centralized data warehouse in the cloud for further processing.

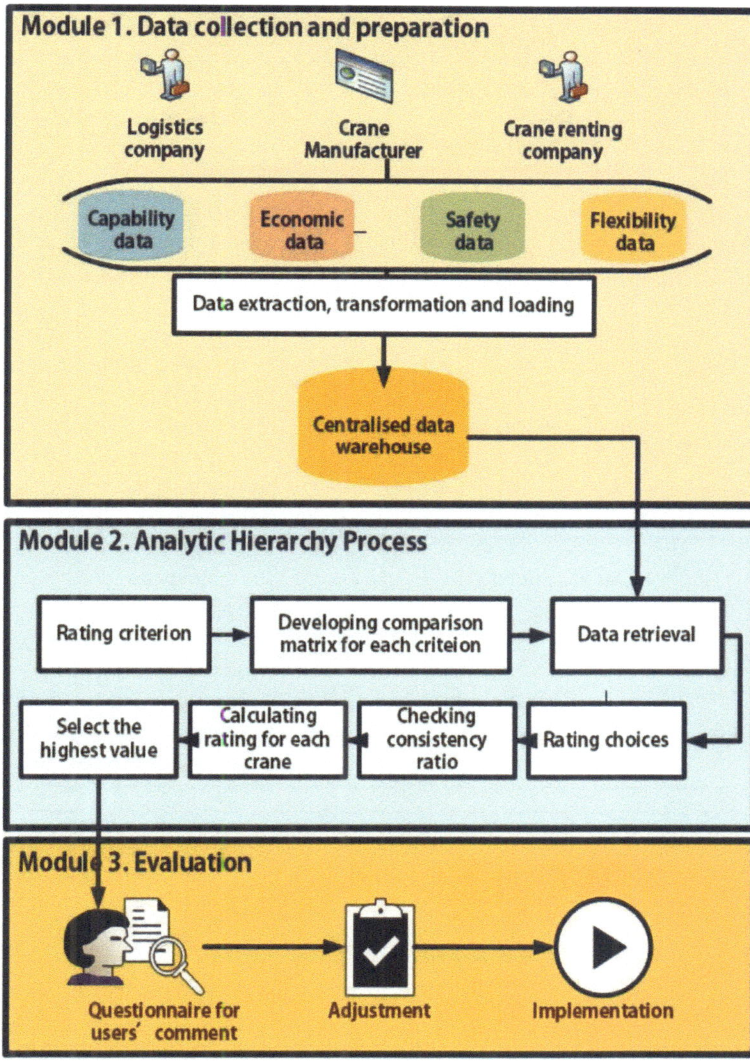

Fig. 8.1 The architecture of the proposed SCSS.

8.2.2 Module 2: Analytic hierarchy process

In module 2, AHP is adopted for the crane selection, which assists the multicriteria decision making of workers by defining different criteria, priority from each criterion and priority in each hierarchy. There are three main parts in applying the procedure in the AHP application: hierarchy construction, pairwise comparisons, and priority weighting.

Table 8.1 Example of data collected for mobile crane selection.

Category	Subcategory	Sample data
Economic data	Renting cost	Rental cost
		Preorder time
	Set-up cost	Set-up time
		Labor cost
		Accessary cost
Safety data	Crane condition	Age of cranes
		Repair history of cranes
		Materials
	Reputation	User's review
		Past history of model
		Manufacturer
		Place of manufacturing
Capability data	Operation	Lifting weight
		Lifting height
		Accuracy
		Turning jib
	Speed	Start-up speed
		Lifting speed
		Buffering speed
		Finishing speed
	Crane relocation	Relocation distance availability
Flexibility data	Complexity	Ability in dealing with barriers
	Duration	Renting duration availability

8.2.2.1 Hierarchy construction

By reviewing the literature and the interviews with experienced workers, the important criteria which affect the decision judgment can be identified for constructing the hierarchy structure. Typically, there are three levels involved in the hierarchy structure: (i) goals defined in the top-level, (ii) criteria and subcriteria considered in the crane selection in the middle level, and (iii) alternatives in the bottom level. In the crane selection, four main criteria, i.e., economic, safety, capability and flexibility, are considered in the proposed SASS, as shown in Fig. 8.2 (Wu et al., 2011). In addition, subcriteria for each criterion are defined. For example, there are three subcriteria for capacity criteria: lifting, engine, and driving speed considered in the crane selection. Moreover, four subcriteria, i.e., jib, loading, boom and radius, are considered when considering the lifting subcriteria.

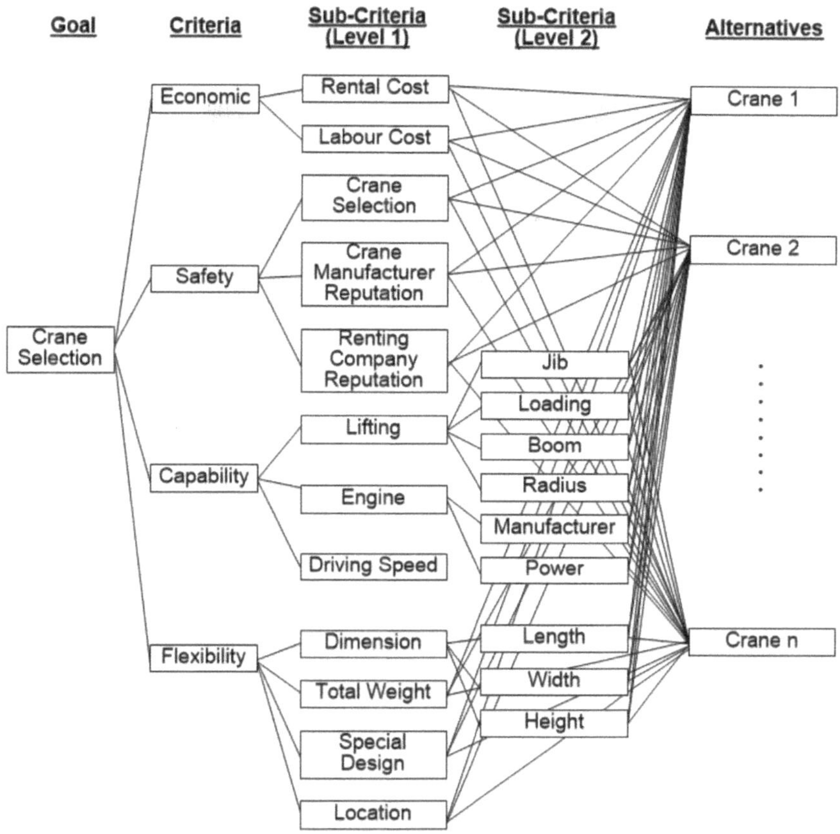

Fig. 8.2 Hierarchy structure for the mobile crane selection in the proposed SASS.

8.2.2.2 Pairwise comparison

After identifying the criteria in each level, pairwise can be compared between each criterion in the same level. Three main steps are involved in the pairwise comparisons as follows:

Step 1: Build a comparison matrix for each level.

Firstly, a 9-point scale is used to judge the relative importance of each pair. The scale is set with precise specifications to maintain consistency in the entire selection, as shown in Table 8.2. After that, a questionnaire is designed and distributed to experienced workers and experts. It is used to quantify the importance of criteria and their subcriteria in the mobile crane selection.

Step 2: Calculate the relative weights for each level.

With the importance of the criteria and its subcriteria, the weighting in each layer is then computed according to the basis of Saaty's eigenvector

Table 8.2 The scale with specification.

Weight intensity	Meaning in comparing criterion	Meaning in comparing attribute
1	Equally important	Equally preferable
3	Slightly more important	Slightly more preferable
5	Essentially more important	Essentially more preferable
7	Greatly more important	Greatly more preferable
9	Absolutely more important	Absolutely more preferable
2,4,6,8	Intermediate level between the adjacent judgment	Intermediate level between the adjacent judgment

procedure. The results of the pairwise comparison are presented in a matrix format as expressed in Eq. (8.1).

$$A = [\alpha_{ij}] = \begin{bmatrix} \alpha_{11} & a_{12} & \cdots & a_{1n} \\ a_{21} & \alpha_{22} & \cdots & a_{2n} \\ \vdots & \vdots & \ddots & \vdots \\ a_{y1} & a_{y2} & \cdots & \alpha_{yn} \end{bmatrix}, \alpha_{ii} = 1, \alpha_{ji} = \frac{1}{\alpha_{ij}}, \alpha_{ij} \neq 0 \tag{8.1}$$

where α_{ij} is the quotient of weights of the criteria, $i = (1, 2, \ldots, y)$ and $j = (1, 2, \ldots, n)$.

Based on the value of α_{ij}, the priority vector W_i can be calculated using Eq. (8.2).

$$W_i = \frac{\sum_{j=1}^{n} a_{ij}}{y}, \forall i = (1, 2, \ldots, y), j = (1, 2, \ldots, n) \tag{8.2}$$

Step 3: Evaluate the consistency of priorities.

It is crucial to ensure a certain level of consistency as it can affect the reliability of the results and keep the plan logically. However, inconsistency cannot be eliminated as it is a process of human judgment. A small degree of inconsistency can be maintained at an acceptable level. The consistency index (CI) is calculated using Eq. (8.3) to address the consistency.

$$\mathrm{CI} = \frac{\sum_{i=1}^{y} \left(\sum_{i=1}^{y} \alpha_{ij} \right) \cdot W_i - y}{y - 1} \tag{8.3}$$

After that, the consistency ratio (CR) can be calculated by dividing the CI with the random consistency index. If the CR is greater than 0.1, it implies the inconsistency that occurred in the α_{ij}. By repeating steps 2 and 3, each criterion's importance and subcriteria can be evaluated.

8.2.3 Module 3: Evaluation

The results from module 2 are provided for the workers to reduce the complexity and improve the efficiency and effectiveness in the mobile crane selection. A case study is conducted and described in Section 8.3 to validate the feasibility. The survey is then conducted in the case company to evaluate the system performance regarding workflow complexity, process time and decision effectiveness.

8.3 Case study

To validate the feasibility of the proposed SASS, a case study is conducted in a medium-sized logistics company in Hong Kong named ABC Limited (Alias). ABC Limited is a freight forwarding company with more than 100 carries in both sea and air freight. Due to the size and position of the company in the market, it aims at providing customized services and seamless value-added services to its customers. It targets eight diverse industries, including aviation, automobile, chemical, healthcare, project cargo and technology, retail and fashion. Project cargo is one of the critical businesses in ABC Limited that provides flexible solutions and professional care to customers projects. Although ABC Limited has the experience and advanced networking in project cargo, it is time-consuming for workers to handle the operation of oversized cargoes and the selection process for the proper crane, which is inefficient. Hence, ABC Limited decides to implement the proposed SASS to improve the reliability of the mobile crane selection process. The implementation procedures of SASS is shown in Fig. 8.3. Five steps are involved: (i) Build up a database, (ii) Set up criterion, (iii) Interview, (iv) System application, and (v) Result and plan.

8.3.1 Build-up database

Firstly, all relevant data are collected to construct the cloud-based centralized database. Nine data types are involved in the centralized database, including the order data, customer data, air waybill data, insurance data, staff data, custom declaration data, truck data, sea waybill data, and crane data. Fig. 8.4 shows the structure of the cloud-based centralized database. These data are used for workers to identify the key criteria and subcriteria for selecting the appropriate crane.

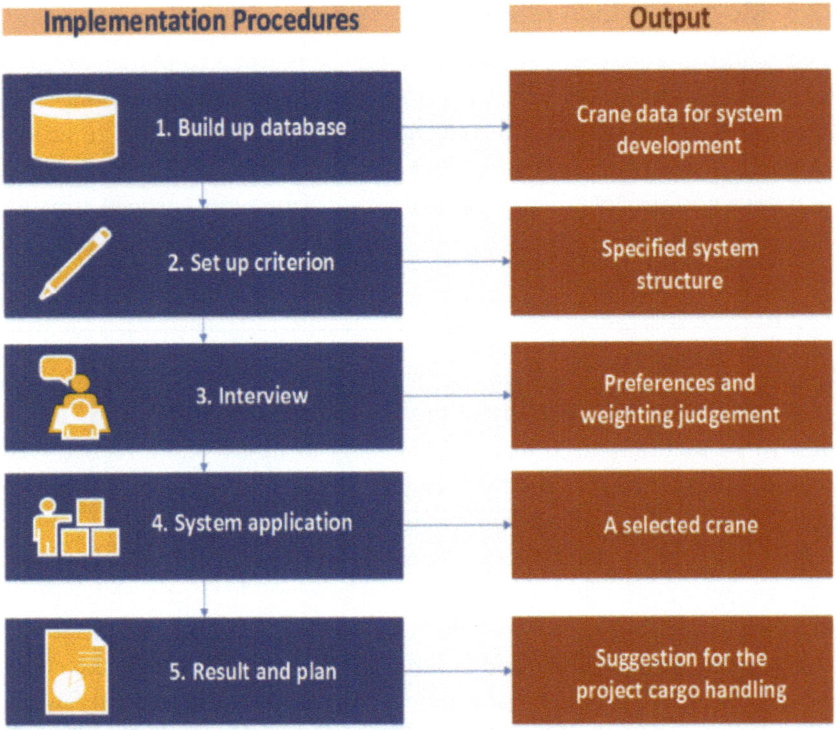

Fig. 8.3 Implementation procedures of SASS.

8.3.2 Set up criterion

In this step, key criteria and their subcriteria for the mobile crane selection are defined for constructing the hierarchy structure of AHP. According to the information mentioned in Fig. 8.2, four criteria and their subcriteria are selected, which may affect the selection of cranes for project cargo. In addition, 15 most frequently used cranes from 6 manufacturers covering capacity ranging from 13 to 300 tons are selected as the alternatives. Table 8.3 shows the detailed information of the selected cranes.

8.3.3 Interview

An interview was conducted to understand the needs of the company in handling various project cargo orders. In this case, the illustration for handling 12 mass objects with a total of 240 tons is applied to show how the AHP can assist workers in selecting mobile cranes. Through the discussion, the preferences of ABC Limited toward each criterion and subcriteria were collected and shown as below:

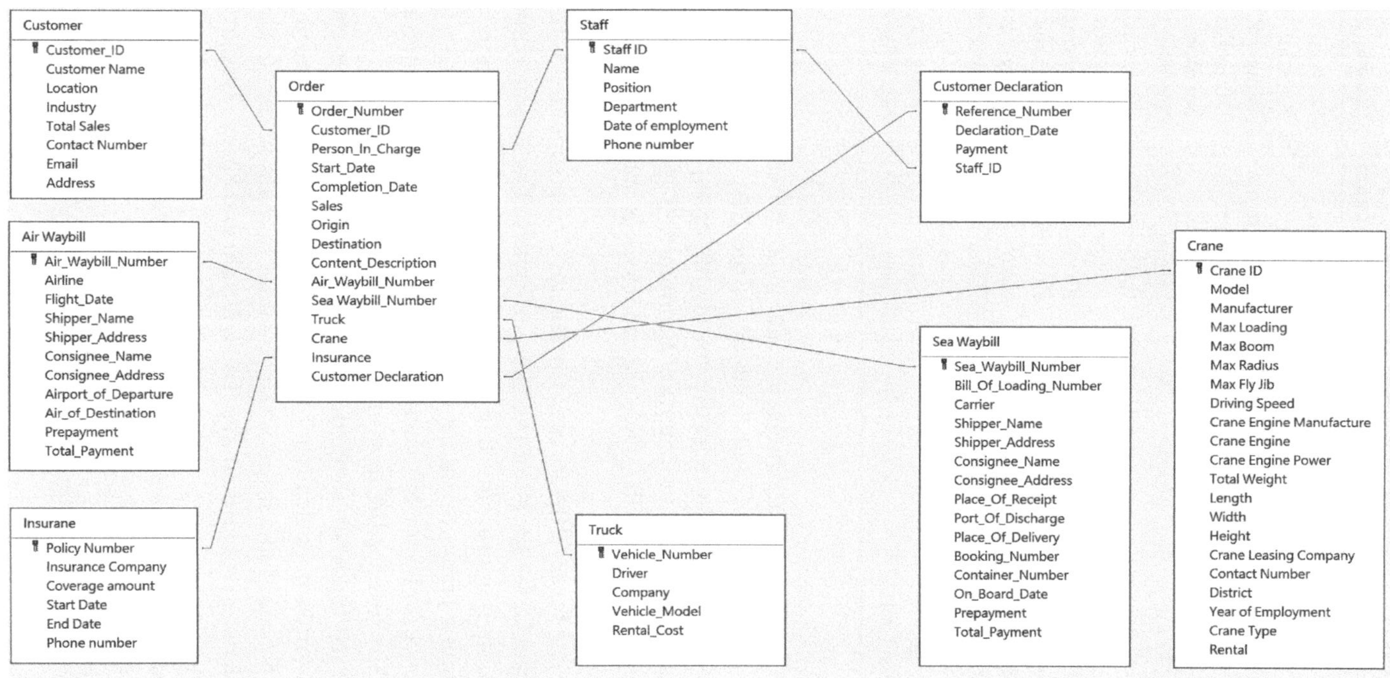

Fig. 8.4 Structure of the cloud-based centralized database.

Table 8.3 Detailed information of the selected cranes.

Item	Model	Capacity (tons)	Manufacturer	Leasing company
1	LTM 1300-6.2	300	Liebherr	Kanson
2	AC 250-1	250	Demag	Kanson
3	LTM 1160/2	160	Liebherr	Wing Fat
4	LTM 1100-4.2	100	Liebherr	Wing Fat
5	SA 1000	100	Sumitomo	Set Win
6	AC70 City	70	Demag	Set Win
7	SL-650R	65	Kato	Towermark
8	LTM 1060-3.1	60	Liebherr	Wing Fat
9	KR50H	50	Kato	Towermark
10	QY50H-2	50	Zoomlion	Chimkee
11	KR 45H-V	45	Kato	Wing Fat
12	GR-500EXL	56	Tadano	Kanson
13	GR-300EX	30	Tadano	Set Win
14	SR 250	25	Kato	Wing Fat
15	MR-130R	13	Kato	Set Win

i. Preference for criteria: Among the four major criteria, a capability is the most crucial criterion, and flexibility is the least important criterion, while Safety and Economics are equally important.

ii. Preference for level 1 subcriteria of Economics: Typically, the rental cost for cranes is about 80%, while the labor cost is about 20% of the total crane cost.

iii. Preference for level 1 subcriteria of Safety: Crane condition is the most significant index to reflect the safety of cranes. The leasing companies' reputation is more critical than crane manufacturers, which implies maintaining cranes to ensure safety.

iv. Preference for level 1 subcriteria of Capability: Lifting is the most important while speed is the least important.

v. Preference for level 1 subcriteria of Flexibility: Location is the most critical criterion in flexibility that the closer location can speed up the entire process. Dimension is the second important in the flexibility to affect the convenience on site. Total Weight and Special Design are less important.

By doing so, the preference for level 2 subcriteria can be distinguished for defining its weighting for AHP calculation.

1. Economics	9	8	7	6	5	4	3	2	**(1)**	2	3	4	5	6	7	8	9	**Safety**
2. Economics	9	8	7	6	5	4	3	2	1	2	3	4	5	6	**(7)**	8	9	**Capability**
3. Economics	9	8	7	6	5	4	**(3)**	2	1	2	3	4	5	6	7	8	9	**Flexibility**
4. Safety	9	8	7	6	5	4	3	2	1	2	3	4	5	6	**(7)**	8	9	**Capability**
5. Safety	9	8	7	6	5	4	**(3)**	2	1	2	3	4	5	6	7	8	9	**Flexibility**
6. Capability	**(9)**	8	7	6	5	4	3	2	1	2	3	4	5	6	7	8	9	**Flexibility**

Fig. 8.5 Pairwise results for the four major criteria in level 2.

8.3.4 System application

In this step, the pairwise comparison is then applied to quantify each criterion's preference and subcriteria using a 9-point scale. Fig. 8.5 shows the pairwise results for the four major criteria.

With the pairwise results, the relative importance can be denoted in matrix $A_{\text{level 2}}$ as shown below:

$$A_{\text{level 2}} = \begin{bmatrix} 1 & 1 & 1/7 & 3 \\ 1 & 1 & 1/7 & 3 \\ 7 & 7 & 1 & 9 \\ 1/3 & 1/3 & 1/9 & 1 \end{bmatrix}$$

After that, $A_{\text{level 2}}$ is normalized for calculating the weightings for each criterion using Eq. (8.2). Table 8.4 shows the relative importance and priority factor for each criterion.

$$W_{\text{level 2}} = \frac{1}{4}\begin{bmatrix} 3/28 & 3/28 & 9/88 & 3/16 \\ 3/28 & 3/28 & 9/88 & 3/16 \\ 3/4 & 3/4 & 63/88 & 9/16 \\ 1/28 & 1/28 & 7/88 & 1/16 \end{bmatrix} = \begin{bmatrix} 0.126 \\ 0.126 \\ 0.695 \\ 0.053 \end{bmatrix}$$

Table 8.4 Relative importance and priority factor for the four major criteria.

Criteria	Economic	Safety	Capability	Flexibility	Priority factor
Economic	1	1	1/7	3	0.126
Safety	1	1	1/7	3	0.126
Capability	7	7	1	9	0.695
Flexibility	1/3	1/3	1/9	1	0.053
Sum	28/3	28/3	88/63	16	100%

Table 8.5 Weighting priority and CR for the subcriteria in level 1.

Criteria	Subcriteria in level 1	Weighting priority	CR
Economic	Labor cost	0.167	0.000
	Rental cost	0.833	
Safety	Crane condition	0.637	0.037
	Crane manufacturer's reputation	0.258	
	Leasing company's reputation	0.105	
Capability	Engine	0.178	0.028
	Lifting	0.751	
	Speed	0.071	
Flexibility	Dimension	0.283	0.028
	Total weight	0.583	
	Special design	0.067	
	Location	0.067	

After calculating the priority factor, evaluation of the consistency ratio is required to ensure consistency pairwise. By using Eq. (8.3), the value of CI is 0.031, and CR is 0.034 can be found. Since the value of CR is smaller than 0.1, the pairwise for the four major criteria is acceptable. By repeating the calculation pairwise for each subcriterion in level 1 and level 2, and the selected crane alternatives, the priority and CR can be distinguished, as shown in Tables 8.5–8.6.

8.3.5 Results and plan

Based on the results generated by AHP, the best crane selection for handling the 12 mass objects with a total of 240 tons is found, as shown in Fig. 8.6. The results show that crane 10 with 0.120 has the highest scores among the

Table 8.6 Weighting priority and CR for the subcriteria in level 2.

Subcriteria in level 1	Subcriteria in level 2	Weighting priority	CR
Lifting	Jib	0.200	0.016
	Max loading	0.522	
	Max boom	0.078	
	Max radius	0.200	
Engine	Manufacturer	0.167	0.000
	Power	0.833	
Weighting	Length	0.072	0.062
	Width	0.279	
	Height	0.649	

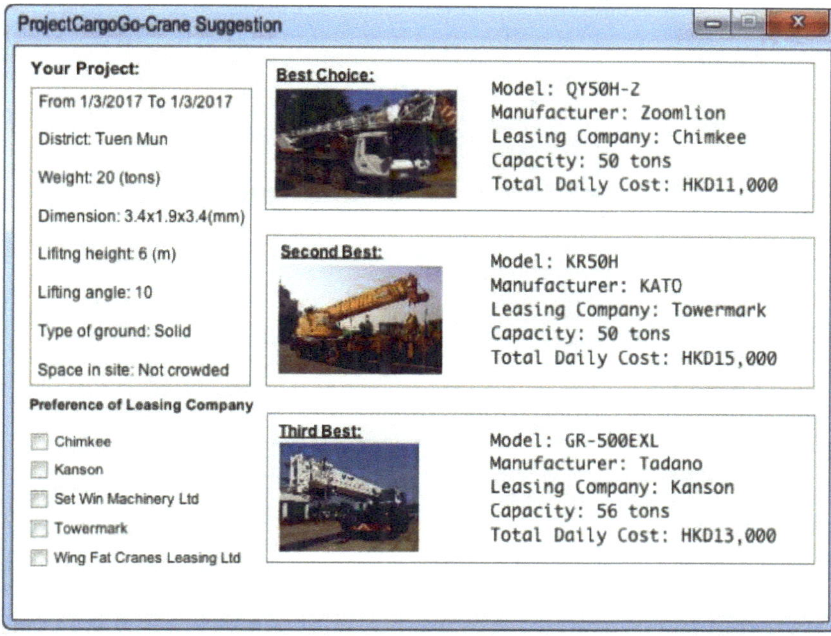

Fig. 8.6 User interface of the proposed SASS to show the selection results.

other crane alternatives. Therefore, crane 10 with 50 tons maximum loading is selected in this case. Workers can effectively plan to book the 10th crane (C10) to complete that project cargo order with such information (Table 8.7).

8.4 Discussion

This chapter proposes a smart approach using the proposed SASS for facilitating the mobile crane selection. Through the pilot study in the ABC Limited, the SASS improves (i) the efficiency and effectiveness of the crane selection and (ii) the service quality in handling project cargo orders.

8.4.1 Improvement in the efficiency and effectiveness of the crane selection

Before implementing the proposed system, the workers in ABC Limited searched three crane leasing companies for suggesting the potential crane for handling the project cargo order according to its nature. With the use of the proposed SASS, the workflow for the crane selection is changed.

Table 8.7 Weight priority of the selected crane alternatives.

	RC	LC	CC	LR	CR	JI	LD	BO	RA	EM	PO	SP	H
C1	0.011	0.012	0.017	0.050	0.015	0.008	0.018	0.011	0.013	0.149	0.192	0.150	0.059
C2	0.013	0.016	0.042	0.050	0.024	0.008	0.018	0.011	0.019	0.075	0.192	0.150	0.012
C3	0.014	0.028	0.050	0.107	0.032	0.033	0.018	0.022	0.047	0.164	0.131	0.059	0.012
C4	0.015	0.079	0.049	0.107	0.158	0.020	0.018	0.022	0.013	0.164	0.102	0.150	0.012
C5	0.017	0.079	0.010	0.015	0.024	0.103	0.018	0.032	0.033	0.015	0.017	0.059	0.025
C6	0.033	0.079	0.010	0.015	0.024	0.103	0.062	0.032	0.046	0.075	0.070	0.108	0.025
C7	0.062	0.079	0.165	0.112	0.070	0.103	0.083	0.049	0.025	0.033	0.040	0.017	0.059
C8	0.026	0.079	0.043	0.112	0.158	0.103	0.113	0.032	0.213	0.134	0.040	0.152	0.039
C9	0.054	0.079	0.010	0.112	0.070	0.103	0.210	0.071	0.049	0.033	0.040	0.015	0.039
C10	0.108	0.079	0.130	0.009	0.010	0.103	0.210	0.071	0.259	0.008	0.040	0.052	0.089
C11	0.090	0.079	0.123	0.116	0.070	0.038	0.010	0.071	0.025	0.033	0.040	0.018	0.041
C12	0.077	0.079	0.049	0.050	0.034	0.089	0.201	0.071	0.105	0.033	0.027	0.018	0.028
C13	0.140	0.079	0.243	0.016	0.034	0.089	0.008	0.136	0.025	0.023	0.017	0.018	0.121
C14	0.152	0.079	0.027	0.112	0.070	0.089	0.008	0.136	0.103	0.031	0.027	0.018	0.121
C15	0.188	0.079	0.031	0.015	0.070	0.008	0.008	0.230	0.025	0.031	0.026	0.018	0.319

RC, rental cost; *LC*, labor cost; *CC*, crane condition; *LR*, leasing company's reputation; *CR*, crane manufacturer's reputation; *JI*, Jib; *LD*, loading; *BO*, boom; *RA*, radius; *EM*, engine manufacturer; *PO*, power of engine; *SP*, speed; *H*, height.

Table 8.8 Improvement in the efficiency and effectiveness of the crane selection.

Tasks in the crane selection	Manual approach	After implementing the SASS	Percentage (%)
Searching crane leasing companies	25 min	0 min	100
Receiving the quotation	8 h	2.5 h	68.75
Deciding the most suitable crane	18 min	5 min	72.22

Workers no longer need to spend time searching for crane leasing companies. They can directly input the special requirements into the system and obtain the most suitable crane for handling the individual project. Therefore, the processing time for selecting cranes can be significantly reduced, as shown in Table 8.8.

8.4.2 Improvement in the service quality in ABC Limited

With the implementation of the proposed SASS, workers can use such knowledge to facilitate decisions and increase the reliability in the mobile crane selection. The SASS also allows the workers to gather the crane information from different crane leasing companies. The service quality in ABC Limited has been improved. After conducting the survey, it is found that the number of complaints due to cargo damage and poor efficiency was significantly decreased by 58.7%. In addition, the service satisfaction of ABC Limited was increased from 50.9% to 85.3%.

8.5 Conclusion

In the trend of globalization and digitalisation, the role of the logistic industry has been raising its significance in supporting the connection around the world. Adopting advanced technologies and artificial intelligence techniques is crucial to improving operational efficiency and effectiveness in the logistics industry. Unlike general cargo, the requirements and procedures for project cargo are more complicated and time-consuming for selecting the proper tools and machines. Due to special attention required for project cargo, improper handling of such project cargo may lead to cargo damage and loss to the company. Therefore, this chapter proposes the smart crane selection system (SASS) integrated data management and

AHP to identify the key factors in selecting mobile cranes and providing decision supports for workers. The proposed SASS had been developed and successfully implemented in ABC Limited. It helps narrow the research gap by presenting the successful, intelligent system in eliminating time of handling, enriching the options of cranes and helping the logistics company gain bargaining power toward the crane leasing companies. However, there are limitations to the proposed system. Firstly, a current approach still relies on the knowledge of experienced workers in identifying the criteria in the crane selection. Future work may focus on the integration of more artificial intelligence techniques and knowledge-based methods to learn and retain the workers' behavior.

References

Delfmann, W., Ten Hompel, M., Kersten, W., Schmidt, T., & Stölzle, W. (2018). Logistics as a science: Central research questions in the era of the fourth industrial revolution. *Logistics Research*, *11*(9), 1–13.

Fang, Y., Cho, Y. K., & Chen, J. (2016). A framework for real-time pro-active safety assistance for mobile crane lifting operations. *Automation in Construction*, *72*, 367–379.

Grossman, G. M., & Helpman, E. (2005). Outsourcing in a global economy. *The Review of Economic Studies*, *72*(1), 135–159.

Lindstad, H., Asbjørnslett, B. E., & Strømman, A. H. (2016). Opportunities for increased profit and reduced cost and emissions by service differentiation within container liner shipping. *Maritime Policy & Management*, *43*(3), 280–294.

Melnyk, O., & Malaksiano, M. (2020). Effectiveness assessment of non-specialised vessel acquisition and operation projects, considering their suitability for oversized cargo transportation. *Transactions on Maritime Science*, *9*(01), 23–34.

Rai, A., Pavlou, P. A., Im, G., & Du, S. (2012). Interfirm IT capability profiles and communications for cocreating relational value: Evidence from the logistics industry. *MIS Quarterly*, *36*(1), 233–262.

Şakar, G. D., Yildirim, E., & Mansuroğlu, E. (2018). Value creation in project cargo logistics: A Delphi study. *Journal of ETA Maritime Science*, *6*(3), 255–274.

UNCTAD. (2015). *Review of maritime transport*. Geneva: UNCTAD.

Wu, D., Lin, Y., Wang, X., Wang, X., & Gao, S. (2011). Algorithm of crane selection for heavy lifts. *Journal of Computing in Civil Engineering*, *25*(1), 57–65.

Zhong, D., Lv, H., Han, J., & Wei, Q. (2014). A practical application combining wireless sensor networks and internet of things: Safety management system for tower crane groups. *Sensors*, *14*(8), 13794–13814.

CHAPTER NINE

Transport, mobility, and delivery in smart cities: The vision of the TransAnalytics research project

9.1 Introduction

Transport and citizens' mobility (T&M) activities represent a key sector in the worldwide economies, contributing to social and economic progress in modern societies. T&M activities also constitute an important source of energy consumption, as well as environmental and social issues. In the face of sharp urbanization worldwide, metropolitan areas have started different initiatives and projects to make cities more energy-efficient and environmentally sustainable. At this moment, T&M activities have a significant impact on the development of the so-called smart cities, where large quantities of data are gathered in real-time via IoT devices (e.g., sensors, cameras, etc.) and analyzed through Cloud systems (Beneicke, Juan, Xhafa, Lopez-Lopez, & Freixes, 2020). The emergence of the on-demand economy (services) and e-commerce (products) has boosted T&M operations in urban and metropolitan areas. Zero-emission vehicles (electric cars, bicycles, etc.) are increasing their presence in our cities. Likewise, new mobility modes, like carsharing and ridesharing, are becoming increasingly popular. Consequently, strategic, planning and operational challenges need to be faced (Juan et al., 2016). For instance, smart cities are required to provide

☆ Research Project, "*Efficient & Sustainable Transport Systems in Smart Cities: Internet of Things, Transport Analytics, and Agile Algorithms*" (*TransAnalytics*) PID2019-111100RB-C21/AEI/ 10.13039/501100011033, Ministerio de Ciencia e Innovación, Spain.

☆☆ With the collaboration of Prof. Angel A. Juan, Project Leader, Open University of Catalonia, Barcelona, Spain.

https://doi.org/10.1016/B978-0-12-820139-8.00004-8

charge stations for electric-based vehicles and parking stations for carsharing mobility. Hence, costly decisions need to be made about these facilities' number, location, and dynamic capacity.

Similarly, some green vehicles' limited driving range and loading capabilities impose additional constraints when designing efficient distribution routes (Eskandarpour et al., 2019; Juan, Goentzel, & Bektaş, 2014). Likewise, ridesharing modes and autonomous vehicles require a new family of "agile" algorithms able to provide efficient routing plans in real-time (a few milliseconds) and re-compute these plans as new data on the traffic status is obtained and processed. Crucial to these algorithms is the use in IoT input, which is computed by IoT data gathered in the city about traffic state. The analytics algorithms that comprise cognitive, descriptive, predictive, or prescriptive analytics from IoT data will feed the T&M system with continuously optimized and contextualized knowledge about system changes and prediction with a high degree of confidence.

This chapter presents the vision of the research project TransAnalytics: "*Efficient & Sustainable Transport Systems in Smart Cities: Internet of Things, Transport Analytics, and Agile Algorithms.*"[a] The project aims at developing transport analytics, sustainable models, T&M algorithms, and software solutions that support stakeholders at the industry, local, and national governments when: (a) quantifying the environmental benefits associated with the use of green technologies (Reyes-Rubiano, Calvet, Juan, Faulin, & Bové, 2020); (b) allowing for real-time an efficient usage of green technologies and T&M modes, including ridesharing, carsharing, and cooperation among carriers (Quintero-Araujo et al., 2019); and (c) contributing to save energy and reduce the social impact of T&M activities. The project has settled the following development stages:

- *Descriptive Transport Analytics*: A description based on open data sources and IoT analytics to achieve maximum advantage under the city environment condition.
- *Predictive Transport Analytics and Modeling*: Development of theoretical models to forecast the evolution of critical key performance indicators (KPIs) in T&M activities.

[a] Research Project, "*Efficient & Sustainable Transport Systems in Smart Cities: Internet of Things, Transport Analytics, and Agile Algorithms*" (*TransAnalytics*) PID2019-111100RB-C21/AEI/ 10.13039/501100011033, Ministerio de Ciencia e Innovación, Spain.

- *Prescriptive Transport Analytics*: Design and implementation of intelligent optimization algorithms to provide efficient solutions to the main challenges associated with T&M activities in smart cities.
- *Knowledge transfer* to (i) enterprises, businesses, and industry; (ii) to local and national institutions; and (iii) to society and international dissemination of the results.

The rest of the chapter is organized as follows. In Section 9.2, related work is briefly presented along various research topics in T&M. Section 9.3 presents the vision of the TransAnalytics project about the research problems and challenges in T&M. In Section 9.4 develops the vision of the project about the role of transport analytics in Smart City T&M. Section 9.5 concludes the chapter.

9.2 Related work

9.2.1 Optimization, metaheuristics, and simulation methods

Optimization and simulation methods and techniques—such as agile algorithms, simheuristics, and learnheuristics—are key methods for providing efficient solvers at scale in various strategic, tactical, and operational decisions in problems related to the development of sustainable and intelligent T&M systems in urban areas. By sharing different resources (e.g., customers, vehicles, warehouses, experts, etc.), companies on the same supply chain level hope to improve their utilization and service levels through economies of scale. Its rising importance is supported in recommendations by different governments (Ministerio de Fomento, 2018; Salmon, 2011), outlining this as an opportunity to establish sustainable T&M systems and make participating companies more competitive. Hence, transport analytics and intelligent algorithms become fundamental tools to analyze and quantify the potential benefits of modern T&M practices and concepts, from horizontal cooperation among carriers to emerging mobility modes such as carsharing and ridesharing. Pérez-Bernabeu et al. (2015), Quintero-Araujo et al. (2019), and Serrano et al. (2017) illustrate possible monetary and environmental savings of sharing customers and vehicle depots.

By employing metaheuristics, collaborative cases and their noncollaborative counterparts are compared. OECD (2015) address cooperation in urban last-mile delivery with uncertain (stochastic) demands. Hezarkhani et al. (2016) discuss gain sharing options through joint delivery planning of logistic service providers. Strategies on a vertical supply chain level have also been addressed in the literature. Liu and Lee (2011) suggest a variable neighborhood tabu search in which they consider a VRP with time windows and its combination with inventory control decisions. Bertazzi et al. (2013) propose a hybrid rollout algorithm for the IRP with stochastic demands. Later, Juan, Grasman, et al. (2014) develop a simheuristic approach for the IRP with stochastic demands and stock-outs for larger problem sizes. Belloso, Juan, and Faulin (2019) and Belloso, Juan, Faulin, and Martinez (2017) analyze the use of heuristic-based algorithms to reduce empty backhauls.

9.2.2 IoT analytics and collective intelligence

With the fast development of IoT and cloud technologies, the use of IoT analytics and collective intelligence have become essential building blocks of agile optimization for T&M systems (Bibri, 2018; Xhafa, 2019; Xhafa et al., 2015, 2017). Indeed, IoT analytics can capture the system dynamics that characterize T&M systems (Yu et al., 2019), and feed the optimization methods in real-time. In addition, IoT technologies bring new business models and value to enterprises (Lee, 2019). The consideration of environmental costs in vehicle routing is a concept that has been evolving over time (Figliozzi, 2010, 2017; Figliozzi et al., 2019; Coelho et al., 2015; Sawik et al., 2017). However, as Demir et al. (2015) noticed, new methodologies to estimate environmental costs are needed. Some of the more efficient ones are the willingness-to-pay estimation (Haddak et al., 2016; Lera-Lopez et al., 2012, 2013) and the experimental economics techniques (Breidert et al., 2006; Croson & Gächter, 2009; Denant-Boèmont et al., 2018).

9.2.3 New transportation means

The increasing use of alternative means of transportation has been supported by research on related problems. The inclusion of electric vehicles in urban transportation fleets and the strategic location of charging stations gave rise to various new VRPs and facility location problems (FLPs) (Juan et al., 2016; Pelletier et al., 2016). Concerning the fleet size and mix, Schneider et al. (2014) model the electric VRP with time windows and recharging stations. The VRP with multiple driving ranges is formally introduced and solved using a multiround heuristic by Juan, Grasman, et al. (2014) and later by Eskandarpour et al. (2019). For the FLP regarding battery recharging stations, You and Hsieh (2014) propose a hybrid heuristic for optimal locations concerning origin-destination trips and alternative station types. Another interesting way of doing urban delivery of goods is the use of bicycles. Using bicycles has recently been revealed as very effective from an environmental point of view (Bagloe et al., 2016; Tipagornwong & Figliozzi, 2014), providing a new paradigm of urban distribution (Bigazzi & Figliozzi, 2012).

9.2.4 Uncertainty in T&M

One of the main issues in modern T&M systems arises through the uncertainty, leading to stochastic and dynamic problem settings (Guimarans, Dominguez, Panadero, & Juan, 2018; Gonzalez et al., 2018; Latorre-Biel et al., 2017). In particular, biased-randomized heuristics (Grasas et al., 2017), simheuristics (Juan et al., 2015), and learnheuristics (Calvet et al., 2017) are tools that have been successfully applied to many realistic and large-sized problem settings. These tools address uncertainty and dynamism in optimization problems in a natural way. As city T&M activities become time more complex under dynamics, uncertainty, and large scale, the current state-of-the-art methods cannot fully address the multidimensional complexity of modern T&M systems. It is necessary to develop an integrated and fully-fledged approach based on open data initiatives, transport analytics, information technologies (IoT, 5G, etc.), sustainability models, and agile optimization to go beyond state of the art.

9.3 Context, research problems, and challenges

Nowadays, about 54% of the global population lives in urban areas, producing near 80% of the worldwide gross domestic product (GDP) (Dobbs et al., 2011). The percentage of people living in metropolitan areas is expected to increase sharply up to 66% by 2050 and up to 85% by the next century. In total numbers, the urban population is expected to increase in the next decades. With growing urban population numbers and booming "mega-cities" (with more than 10 million inhabitants) across the globe, the development of efficient and sustainable urban T&M activities is of major importance for municipalities, especially considering the various economic (profit), environmental (planet), and social (people) impacts through freight T&M (Faulin et al., 2018). In Europe, every year, nearly 1% of the European Union's (EU's) GDP is lost due to traffic congestions.

Concerning the impact on the environment, transport (mainly through road) accounted for 27% of total greenhouse gas emissions in the United States. Even though passenger cars and other light-duty vehicles account for most of this percentage (with about 60%), greenhouse gas emissions from medium and heavy-duty trucks (with about 23% of the total) were the second-largest contributor of NOx emissions. Also, emissions related to freight transportation have increased by over 75% since 1990 (US Environmental Protection Agency, 2013). At the international level too, in 2015 United Nations identified the cooperation policy to urban and peri-urban mobility and urgent actions to combat climate change and its impacts as two of their Sustainable Development Goals included in the "2030 Agenda for Sustainable Development." Next to negative externalities through T&M activities on the economy and the environment, health issues through high NOx concentrations and excessive noise exposure from urban road traffic cannot be neglected (EU Environment Agency, 2009). With this background, the need to develop sustainable and efficient T&M models and systems has become imperative. Savelsbergh and van Woensel (2016) summarize that "city logistics is about finding efficient and effective ways to transport goods in urban areas while considering the negative effects on congestion, safety, and environment."

On the one hand, various T&M challenges in modern business models, e.g., e-commerce, same-day deliveries, location-based services, driven by fierce competition, and challenging customer demands, need to be addressed. On the other hand, technological advances related to the use of electric vehicles and new mobility modes—such as carsharing and ridesharing—promise to diminish the negative impacts of urban T&M activities (Beneicke et al., 2020).

In this context, transport data analytics and hybrid intelligent algorithms, such as those based on the combined use of metaheuristics, simulation-optimization, parallel and distributed computing, and machine learning, can support strategic, tactical, and operational decision-making in the real-time or short term when needed. In particular, we propose the use of biased-randomized heuristics (Grasas et al., 2017), simheuristics (Juan et al., 2015), learnheuristics (Calvet et al., 2017), sensor collected intelligence (Xhafa et al., 2017), and agile optimization algorithms (Juan et al., 2019) as an efficient and original way to deal with some of the most complex problems in urban T&M.

9.3.1 Research problems and challenges in city T&M

City T&M is a very broad concept, leading to a wide range of research problems and challenges (Gonzalez-Feliu & Routhier, 2014). For example, alternative technologies such as electric vehicles are increasingly used by many companies as an initiative for cleaner transportation. These technologies aim to reduce the dependence on oil-based energy sources, which will also reduce the carbon footprint. Related research challenges associated with the use of alternative means of transportation include the inclusion of externalities such as the environmental impact of routing in optimization models, problems related to fleet size and mix, the optimal location of charging stations, and resulting vehicle routing problems (VRPs) with multiple driving ranges, heterogeneous vehicles, backhauling issues, and recharging times (Juan et al., 2016).

9.3.2 Business strategies and opportunities in city T&M

A range of new business strategies and supply chain networks also result in many optimization challenges. Especially in urban areas, the development of multiechelon networks is receiving increased attention (Mancini, 2013). Satellite facilities and urban distribution centers at city borders and strategically important points in city centers can be named in this context. The idea is to decrease heavy freight/goods vehicles in city centers to decrease noise/air pollution and avoid inner-city traffic jams caused by T&M activities. Arising research challenges include multiechelon routing problems, last-mile delivery in multimodal supply chains, or joint delivery concepts (Cuda et al., 2015).

Furthermore, companies are increasingly interested in new transportation concepts that allow them to consolidate freight volumes and achieve higher vehicle utilization. A type of cooperation involves different supply chain partners (e.g., suppliers, manufacturers, logistics service providers, and customers) working together. This cooperation leads to problem settings such as the complex inventory routing problem (IRP) in which routing and inventory decisions are combined (Coelho et al., 2013; Gruler et al., 2019a, 2019b). Achieving a higher vehicle utilization is imperative for new vehicle types which account for constrained energy resources (battery). The project will investigate how to increase the efficiency of novel models for zero-emission city mobility, including ridesharing and carsharing (Katzev, 2003) using electric vehicles. Unlike traditional vehicles, there are no known patterns of energy consumption for electric vehicles while moving in a city.

In the area of T&M activities, despite the fact that common technologies and tools are available for enhancing decision-making processes based on empirical evidence, many enterprises are still far from employing the full potential of such state-of-the-art scientific advances in their day-to-day operations operational processes. As a result, efforts are often wasted, opportunities to economize are lost, and citizens end up paying higher bills for products and services. Road transport is the predominant way of transporting goods in Europe and other parts of the world. Direct costs associated with this type of transportation have increased significantly since 2000. Moreover, road transport is intrinsically associated with a good deal of indirect or external costs—congestion, contamination, security and safety-related costs, mobility, delay time costs, etc.—but are usually left unaccounted for because of the difficulty of quantifying them. In addition to these tangible costs, many others—like environmental costs of producing and using fossil fuel—enter the green equation. According to the results in a recent study (Day et al., 2018), urban low-emissions policies that promote the improvement of air quality parameters would prevent 1.3 million premature deaths until 2030.

Moreover, the economic cost associated with mortality from air pollution is enormous, so low emissions will directly help mitigate this impact. The benefits of healthier physical activity, such as walking, running, or cycling in cities with a low-emissions policy, are significant. On the contrary, the adverse effects associated with inhaled air pollution need to be avoided. Likewise, the impact of the reduction of noise levels caused by a large fleet of combustion vehicles cannot be underestimated.

Therefore, to increase competitiveness in a global economy, European enterprises in T&M must benefit from efficient approaches (simulation-optimization models, agile algorithms, expert systems, and real-time information systems) that promote efficient and environmentally sustainable decision-making. The recent debate on the use of diesel-powered vehicles in urban centers is an example of how things are quickly changing toward the use of less polluting technologies. However, from an industrial perspective, the migration toward greener technologies in daily T&M operations must be done efficiently, so it does not represent a significant increase in monetary costs for them. Thus, there are plenty of opportunities to transfer knowledge in this research area to industries in the T&M sector.

9.3.3 Research and development of dynamic delivery systems

Hereby, the dynamic nature of deliveries might refer to dynamic input variables to routing, location, and scheduling problems. The inputs could be customer demands and travel times, which can be transferred and processed in almost real-time through the use of modern information & communication technologies, IoT, efficient & agile optimization algorithms, metaheuristics, and simheuristics (Ritzinger et al., 2016). Moreover, dynamism might also refer to incoming orders. Business models addressing such problems include, for example, crowdshipping and the use of occasional drivers, as implemented by Walmart (Barr & Wohl, 2013).

9.3.4 Analysis of real-life case studies of the T&M and benefits

The area of smart city T&M offers many potential use cases, among which, we distinguish the following:

- *T&M models for electric vehicles and sharing strategies* comprising the use of carsharing and ridesharing, shared parking slots, use of shared consolidation centers, use of shared vehicles and workers, integration of electric vehicles in fleets, etc.; these require both IoT analytics and agile optimization to solve in real-time allocation, location-and-route planning problems as well as profiling of energy consumption of electric vehicles for optimizing their autonomy.
- *T&M models for home-delivery services.* Likewise, this case study requires IoT analytics and agile optimization to solve real-time route planning problems. Finally, this task will identify and classify best practices associated with the development, integration, and implementation of management information systems in the context of smart city T&M activities.
- *T&M Cost Models and Impact.* In the area of T&M activities, despite the fact that common technologies and tools are available for enhancing decision-making processes based on empirical evidence, many enterprises are still far from employing the full potential of such state-of-the-art scientific advances in their day-to-day operations operational processes. As a result, efforts are often wasted, opportunities to economize are lost, and citizens end up paying higher bills for products and services. Moreover, there should be considered the economic cost associated with mortality from air pollution is enormous, so low emissions will directly help to mitigate this impact.

9.4 The role of transport analytics in smart city T&M

The vision of the TransAnalytics project is to develop transport analytics as a basis for sustainable models, T&M algorithms, and software solutions that support stakeholders at the industry, local, and national governments when:

(a) Quantifying the environmental benefits associated with the use of green technologies (Reyes-Rubiano et al., 2020).

(b) Allowing for real-time an efficient usage of green technologies and T&M modes, including ridesharing, carsharing, and cooperation among carriers (Quintero-Araujo et al., 2019).

(c) Contributing to save energy and reduce the social impact of T&M activities.

The building blocks of the transport analytics are as follows next.

9.4.1 Descriptive transport analytics

The computation of descriptive analytics is based on open data sources (such as the *Open Data Barcelona initiative*[b]) and IoT analytics to achieve maximum advantage under the city environment. The IoT analytics will feed the real-time cognitive analytics in input to an agile optimization engine, which will then enable the T&M System to take real-time decisions.

9.4.2 Predictive transport analytics

Predictive Transport Analytics is very useful for developing theoretical models to forecast the evolution of critical key performance indicators (KPIs) in T&M activities and modeling T&M scenarios in smart cities. The use of ridesharing and carsharing mobility modes, consolidation centers, shared vehicles and parking areas, charging stations, heterogeneous fleets composed of traditional and electric vehicles, real-time information systems, energy consumption profiling, etc. are the possible scenarios. To this end, the support by a "data lake" and data warehouse of semantically enriched data from IoT gathered data is fundamental.

9.4.3 Prescriptive transport analytics

Prescriptive Transport Analytics is envisaged as a basis for designing and implementing intelligent optimization algorithms to provide efficient solutions to the main challenges associated with T&M activities in smart cities. Depending on the decision level (strategic, tactical, or operational), these algorithms can consider scenarios with uncertainty (Juan et al., 2018),

[b] Open Data BCN: https://opendata-ajuntament.barcelona.cat/en/open-data-bcn.

dynamics (Arnau et al., 2018), and/or real-time requirements (agile optimization algorithms).

9.4.4 The role of IoT analytics in transport analytics

The TransAnalytics project aims to bridge the gap and integrate the IoT analytics into an intelligent T&M system, namely, the development of an IoT analytics platform and its integration with the T&M optimization module. Analyzing best practices and implementation issues directly related.

To the use of management information systems in T&M, the aims are

- Integration of IoT technologies, Big Data, and information systems in the city T&M models.
- Real-life applications of IoT technologies and information systems for T&M.

To achieve such objectives, IoT Analytics comprises the following three building blocks, described next.

9.4.4.1 IoT network in Smart City

The IoT network in the city is the platform of IoT devices that will dynamically gather data about the traffic status environment. This platform can be made up of

- IoT devices embedded in vehicles, namely On-Board Units (OBUs).
- Mobile devices of users offering GPS like services.
- RSU-Road Side Units and existing IoT devices, such as high-end cameras, sensors, and gateways deployed in the city by the City Hall.
- Gateways, RSUs, and 5G servers in tower-base stations will be in charge of efficient data gathering & aggregation by preprocessing and filtering data before moving it to the Cloud.

9.4.4.2 Data lake and data warehouse of semantically enriched data

At the edges of the Internet, the T&M system needs to maintain a data lake platform to gather and temporarily store raw data gathered by IoT devices. This data will then be preprocessed, filtered, normalized, and analyzed to yield real-time analytics. However, a more significant amount of data can be very useful to get meaningful insights for short and mid-term action. Therefore, a layer for semantically enriching the data with contextual information and storing it in a BigData warehouse at the Cloud level will be sought.

9.4.4.3 Deep cognitive analytics

These analytics are based on semantically enriched data and contribute to achieving higher optimization levels for the T&M system. Such transport analytics go beyond statistical analysis by using machine learning techniques on data streams and historical data to identify patterns, correlations, trends and build predictive models for use in the agile optimization layer.

9.5 Conclusions

In this chapter, we have presented the vision of the research project TransAnalytics: "*Efficient & Sustainable Transport Systems in Smart Cities: Internet of Things, Transport Analytics, and Agile Algorithms*," which aims at developing transport analytics, sustainable models, T&M algorithms, and software solutions that support stakeholders at the industry, local, and national governments. The project is based on four pillars: *Descriptive Transport Analytics*, *Predictive Transport Analytics and Modeling*, *Prescriptive Transport Analytics* and *Knowledge transfer* to enterprises, businesses, and industry, among others. Throughout the chapter, we have analyzed the vision of the project for agile optimization, IoT analytics, and simheuristics to achieve the challenging objectives of T&M in smart cities. The road ahead, issues, and challenges are also discussed in the multidisciplinary field of optimization, IoT, Fog and Cloud technologies, big data, and analytics at large for efficient and sustainable T&M.

Acknowledgment

This work is supported by Research Project, "*Efficient & Sustainable Transport Systems in Smart Cities: Internet of Things, Transport Analytics, and Agile Algorithms*" (TransAnalytics) PID2019-111100RB-C21/AEI/ 10.13039/501100011033, Ministerio de Ciencia e Innovación, Spain.

References

Arnau, Q., Juan, A., & Serra, I. (2018). On the use of learnheuristics in vehicle routing optimisation problems with dynamic inputs. *Algorithms*, *11*(12), 208.

Bagloe, S. A., Sarvi, M., & Wallace, M. (2016). Bicycle lane priority: Promoting bicycle as a green mode even in congested urban area. *Transportation Research Part A*, *87*, 102–121.

Barr, A., & Wohl, J. (2013). Walmart may get customers to deliver packages to online buyers. *Reuters*. https://www.reuters.com/article/us-retail-walmart-delivery-idUSBRE92R03820130328.

Belloso, J., Juan, A., & Faulin, J. (2019). An iterative biased-randomized heuristic for the fleet size and mix vehicle routing problem with backhauls. *International Transactions in Operational Research*, *26*, 289–301. https://doi.org/10.1111/itor.12379.

Belloso, J., Juan, A., Faulin, J., & Martinez, E. (2017). A biased-randomised metaheuristic for the vehicle routing problem with clustered and mixed backhauls. *Networks*, *69*(3), 241–255.

Beneicke, J., Juan, A. A., Xhafa, F., Lopez-Lopez, D., & Freixes, A. (2020). Empowering citizens' cognition and decision making in smart sustainable cities. *IEEE Consumer Electronics Magazine*, *9*(1), 102–108.

Bertazzi, L., Bosco, A., Guerriero, F., & Laganà, D. (2013). A stochastic inventory routing problem with stock-out. *Transportation Research Part C: Emerging Technologies*, *27*, 89–107.

Bibri, S. E. (2018). The IoT for smart sustainable cities of the future: An analytical framework for sensor-based big data applications for environmental sustainability. *Sustainable Cities and Society*, *38*, 230–253.

Bigazzi, A. Y., & Figliozzi, M. A. (2012). Congestion and emissions mitigation: A comparison of capacity, demand, and vehicle based strategies. *Transportation Research Part D*, *17*, 538–547.

Breidert, C., Hahsler, M., & Reuttener, T. (2006). A review of methods for measuring willingness-to-pay. *Innovative Marketing*, *2*, 8–32.

Calvet, L., De Armas, J., Masip, D., & Juan, A. (2017). Learnheuristics: Hybridising metaheuristics with machine learning for optimisation with dynamic inputs. *Open Mathematics*, *15*, 261–280.

Coelho, L. C., Cordeau, J.-F., & Laporte, G. (2013). Thirty years of inventory routing. *Transportation Science*, *48*(1), 1–19.

Coelho, L. C., Renaud, J., & Laporte, G. (2015). Road-based goods transportation: A survey of real-world applications from 2000 to 2015. In *No. CIRRELT-2015-34. Technical Report FSA-2015-007, Québec, Canada, 2015*.

Croson, R., & Gächter, S. (2009). The science of experimental economics. *Journal of Economic Behavior and Organization*, *73*(1), 122–151.

Cuda, R., Guastaroba, G., & Speranza, M. G. (2015). A survey on two-echelon routing problems. *Computers and Operations Research*, *55*, 185–199.

Day, T., Gonzales-Zuñiga, S., Nascimento, L., Höhne, N., Fekete, H., Sterl, S., & Chen, W. (2018). *Climate opportunity: More jobs; better health; liveable cities*. https://newclimate.org/wp-content/uploads/2018/09/ClimateOpportunity_Methodologies.pdf.

Demir, E., Yuan, H., Scholts, S., & Van Woensel, T. (2015). A selected review on the negative externalities of the freight transportation: Modeling and pricing. *Transportation Research Part E*, *77*, 95–114.

Denant-Boèmont, L., Faulin, J., Hammiche, S., & Serrano-Hernandez, A. (2018). Managing transportation externalities in the Pyrenees region: Measuring the willingness-to-pay for road freight noise reduction using an experimental auction mechanism. *Journal of Cleaner Production*, *202*, 631–641.

Dobbs, R., Smit, S., Remes, J., Manyika, J., Roxburgh, C., & Restrepo, A. (2011). *Urban world: Mapping the economic power of cities (McKinsey Global Institute)*. https://www.mckinsey.com/featured-insights/urbanization/urban-world-mapping-the-economic-power-of-cities.

Eskandarpour, M., Ouelhadj, D., Hatami, S., Juan, A. A., & Khosravi, B. (2019). Enhanced multi-directional local search for the bi-objective heterogeneous vehicle routing problem with multiple driving ranges. *European Journal of Operational Research*, *277*(2), 479–491.

European Environment Agency. (2009). *EEA draws the first map of Europe's noise exposure*. http://www.eea.europa.eu/media/newsreleases/eea-draws-the-first-map-of-europe2019s-noise-exposure.

Faulin, J., Grasman, S., Juan, A., & Hirsch, P. (Eds.). (2018). *Sustainable transportation and smart logistics: Decision making models and solutions* Elsevier, ISBN:9780128142424.

Figliozzi, M. A. (2010). *Vehicle routing problem for emissions minimization.* Transportation Research Record: Journal of the Transportation Research Board, No. 2197 (pp. 1–7). Washington, DC: Transportation Research Board of the National Academies. https://doi.org/10.3141/2197-01.

Figliozzi, M. A. (2017). Lifecycle modeling and assessment of unmanned aerial vehicles (drones) CO2e emissions. *Transportation Research Part D: Transport and Environment, 57*, 251–261.

Figliozzi, M. A., Saenz, J., & Faulin, J. (2019). Minimisation of urban freight distribution lifecycle CO2e emissions: Results from an optimisation model and a real-world case study. *Transport Policy, 86*(February 2020), 60–68.

Gonzalez, S., Juan, A., Riera, D., Elizondo, M., & Ramos, J. (2018). A Simheuristic algorithm for solving the arc routing problem with stochastic demands. *Journal of Simulation, 12*(1), 53–66.

Gonzalez-Feliu, J., & Routhier, J. (2014). *Sustainable urban logistics: Concepts, methods and information systems.* Springer.

Grasas, A., Juan, A., Faulin, J., De Armas, J., & Ramalhinho, H. (2017). Biased randomisation of heuristics using skewed probability distributions: A survey and some applications. *Computers and Industrial Engineering, 110*, 216–228.

Gruler, A., Panadero, J., de Armas, J., Moreno, A., & Juan, A. (2019). A variable neighborhood search based simheuristic for the inventory routing problem with stochastic demands. *Computers and Industrial Engineering, 123*, 278–288.

Gruler, A., Panadero, J., de Armas, J., Moreno, J., & Juan, A. (2020). A variable neighborhood search simheuristic for the multi-period inventory routing problem with stochastic demands. *International Transactions in Operational Research, 27*(1), 314–335. https://doi.org/10.1111/itor.12540.

Guimarans, D., Dominguez, O., Panadero, J., & Juan, A. A. (2018). A simheuristic approach for the two-dimensional vehicle routing problem with stochastic travel times. *Simulation Modelling Practice and Theory, 89*, 1–14.

Haddak, M. M., Lefèvre, M., & Havet, N. (2016). Willingness-to-pay for road improvement. *Transportation Research Part A, 87*, 1–10.

Hezarkhani, B., Slikker, M., & van Woensel, T. (2016). A competitive solution for cooperative truckload delivery. *OR Spectrum, 83*, 51–80.

Juan, A. A., Faulin, J., Grasman, S. E., Rabe, M., & Figueira, G. (2015). A review of simheuristics: Extending metaheuristics to deal with stochastic combinatorial optimisation problems. *Operations Research Perspectives, 2*, 62–72.

Juan, A. A., Goentzel, J., & Bektaş, T. (2014). Routing fleets with multiple driving ranges: Is it possible to use greener fleet configurations? *Applied Soft Computing, 21*, 84–94.

Juan, A. A., Grasman, S. E., Caceres-Cruz, J., & Bektaş, T. (2014). A simheuristic algorithm for the single-period stochastic inventory-routing problem with stock-outs. *Simulation Modelling Practice and Theory, 46*, 40–52.

Juan, A. A., Kelton, W. D., Currie, C. S., & Faulin, J. (2018). Simheuristics applications: Dealing with uncertainty in logistics, transportation, and other supply chain areas. In *Proceedings of the 2018 Winter Simulation Conference* (pp. 3048–3059). IEEE Press.

Juan, A. A., Mendez, C., Faulin, J., Armas, J., & Grasman, S. (2016). Electric vehicles in logistics and transportation: A survey on emerging environmental, strategic, and operational challenges. *Energies, 9*(86).

Juan, A. A., Panadero, J., Bayliss, C., Martins, L., Freixes, A., & Raba, D. (2019). Agile optimization in transportation and logistics. In *Proceedings of the XXXVIII Congreso Nacional de Estadística e Investigación Operativa, Alcoy, Spain, September 3-6* (p. 142)., ISBN:978-84-09-13580-6.

Katzev, R. (2003). Car sharing: A new approach to urban transportation problems. *Analyses of Social Issues and Public Policy, 3*(1), 65–86.

Latorre-Biel, J. I., Faulin, J., Jiménez, E., & Juan, A. A. (2017). Simulation model of traffic in smart cities for decision-making support: Case study in Tudela (Navarre, Spain). In *Proceedings of the international conference on smart cities* (pp. 144–153). Springer.

Lee, I. (2019). The Internet of Things for enterprises: An ecosystem, architecture, and IoT service business model. *Internet of Things*, 7(September).

Lera-Lopez, F., Faulin, J., & Sanchez, M. (2012). Determinants of the willingness-to-pay for reducing the environmental impacts of road transportation. *Transportation Research Part D Transport and Environment, 17*, 215–220.

Lera-Lopez, F., Faulin, J., & Sanchez, M. (2013). Willingness to pay to reduce environmental impacts from road transportation: A case study from the Spanish Pyrenees. *Journal of Applied Operational Research, 5*(4), 135–152.

Liu, S.-C., & Lee, W.-T. (2011). A heuristic method for the inventory routing problem with time windows. *Expert Systems with Applications, 38*(10), 13223–13231.

Mancini, S. (2013). Multi-echelon distribution systems in city logistics. *European Transport Trasporti Europei, 54*, 1–24.

Ministerio de Fomento. (2018). *Observatorio de costes del transporte de mercancías por carretera.* https://www.fomento.gob.es/transporte-terrestre/servicios-al-transportista/observatorios-del-transporte/observatorios-del-transporte-de-mercancias-por-carretera.

OECD. (2015). *The metropolitan century: Understanding urbanisation and its consequences.* Paris: OECD Publishing. https://doi.org/10.1787/9789264228733-en.

Pelletier, S., Jabali, O., & Laporte, G. (2016). Goods distribution with electric vehicles: Review and research perspectives. *Transportation Science, 50*, 3–22.

Pérez-Bernabeu, E., Juan, A. A., Faulin, J., & Barrios, B. B. (2015). Horizontal cooperation in road transportation: A case illustrating savings in distances and greenhouse gas emissions. *International Transactions in Operational Research, 22*(3), 585–606.

Quintero-Araujo, C. L., Gruler, A., Juan, A. A., & Faulin, J. (2019). Using horizontal cooperation concepts in integrated routing and facility-location decisions. *International Transactions in Operational Research, 26*(2), 551–576.

Reyes-Rubiano, L., Calvet, L., Juan, A. A., Faulin, J., & Bové, L. (2020). A biased-randomized variable neighborhood search for sustainable multi-depot vehicle routing problems. *Journal of Heuristics, 26*(3), 401–422.

Ritzinger, U., Puchinger, J., & Hartl, R. F. (2016). A survey on dynamic and stochastic vehicle routing problems. *International Journal of Production Research, 54*(1), 215–231.

Salmon, K. (2011). *Pratiques de logistique collaborative: quelles opportunités pour les PME/ETI? Pôleinter ministériel de prospective et d'anticipation des mutations économiques (PIPAME).* http://www.kurtsalmon.com/uploads/Logistique-collaborative-rapport-complet.pdf.

Savelsbergh, M. W. P. M., & van Woensel, T. (2016). City logistics: Challenges and opportunities. *Transportation Science, 50*(2), 579–590. https://doi.org/10.1287/trsc.2016.0675.

Sawik, B., Faulin, J., & Pérez-Bernabeu, E. (2017). Multi-criteria optimisation for fleet size with environmental aspects. *Transportation Research Procedia, 27*, 61–68.

Schneider, M., Stenger, A., & Goeke, D. (2014). The electric vehicle-routing problem with time windows and recharging stations. *Transportation Science, 48*, 500–520.

Serrano, A., Juan, A., Faulin, J., & Perez, E. (2017). Horizontal collaboration in freight transport logistics: Concepts, benefits, and environmental challenges. *Statistics and Operations Research Transactions, 41*(2), 393–414.

Tipagornwong, C., & Figliozzi, M. (2014). *Analysis of competitiveness of freight tricycle delivery services in urban areas* (pp. 76–84). Washington, DC: Transportation research record: Journal of the Transportation Research Board, No. 2410.

United States Environmental Protection Agency. (2013). *Greenhouse gas emissions 1990-2013.* http://www3.epa.gov/otaq/climate/documents/420f15032.pdf.

Xhafa, F. (2019). Special issue "internet of things (IoT)-based services". *Computing, 101*(7), 725–727.

Xhafa, F., Barolli, L., Barolli, A., & Papajorgji, P. (Eds.). (2015). *Modeling and processing for next-generation big-data technologies. With applications and case studies* Springer.

Xhafa, F., Leu, F.-Y., & Hung, L. L. (Eds.). (2017). *Smart sensors networks. Communication technologies and intelligent applications* Academic Press.

You, P. S., & Hsieh, Y. C. (2014). A hybrid heuristic approach for the problem of the location of vehicle charging stations. *Computers & Industrial Engineering, 70*, 195–204.

Yu, Z., Xhafa, F., Chen, Y., & Kun Ma, K. (2019). A distributed hybrid index for processing continuous range queries over moving objects. *Soft Computing, 23*(9), 3191–3205.

Further reading

Bektas, T., Crainic, T. G., & van Woensel, T. (2017). From managing urban freight to smart city logistics networks. In K. Gakis, & P. Pardalos (Eds.), *Network design and optimization for smart cities (pp. 143–188, Series on Computers and Operations Research, Vol. 8)* World Scientific.

De Armas, J., Cadarso, L., Juan, A., & Faulin, J. (2016). A multi-start randomised heuristic for real-life crew rostering problems in airlines with work-balancing goals. *Annals of Operations Research, 258*(2), 825–848.

De Armas, J., Keenan, P., Juan, A. A., & McGarraghy, S. (2019). Solving large-scale time capacitated arc routing problems: From real-time heuristics to metaheuristics. *Annals of Operations Research, 273*(1–2), 135–162.

De Marinis, E., Andreucci, F., Pucci, F., & Uliana, M. (2015). *Contextual processing for pedestrian tracking in GPS-denied environments.* FOSS4G Europe 2015, 14–17 Julio.

Estrada-Moreno, A., Cavero-Lazaro, M., Juan, A. A., Serrat, C., & Nogal, M. (2019). Distribution planning in a weather-dependent scenario with stochastic travel times: A simheuristic approach. In *Proceedings of the 2018 winter simulation conference* (pp. 3013–3024). IEEE Press.

Estrada-Moreno, A., Fikar, C., Juan, A. A., & Hirsch, P. (2019). A biased-randomised algorithm for redistribution of perishable food inventories in supermarket chains. *International Transactions in Operational Research, 26*(6), 2077–2095.

Estrada-Moreno, A., Savelsbergh, M., Juan, A. A., & Panadero, J. (2019). Biased-randomized iterated local search for a multiperiod vehicle routing problem with price discounts for delivery flexibility. *International Transactions in Operational Research, 26*(4), 1293–1314.

Figliozzi, M. A., & Tipagornwong, C. (2017). Impact of last mile parking availability on commercial vehicle costs and operations. *Supply Chain Forum: An International Journal, 18*(2), 60–68.

György, K., Attila, A., & Tamás, F. (2017). New framework for monitoring urban mobility in European cities. 3rd Conference on Sustainable Urban Mobility, 3rd CSUM 2016, Greece. *Transportation Research Procedia, 24*, 155–162.

Juan, A. A., Faulin, J., Jorba, J., Riera, D., Masip, D., & Barrios, B. (2011). On the use of Monte Carlo simulation, cache and splitting techniques to improve the Clarke and Wright savings heuristics. *Journal of the Operational Research Society, 62*(6), 1085–1097.

Juan, A. A., & Rabe, M. (2013). Combining simulation with heuristics to solve stochastic routing and scheduling problems. In W. Dangelmaier, C. Laroque, & A. Klaas (Eds.), *Simulation in produktion und logistik* (pp. 641–649). Paderborn: Heinz-Nixdorf-Institute.

Juan, A. A., Reyes, L., de la Torre, R., Panadero, J., Faulin, J., & Latorre, I. (2020). Simulation-based optimisation in transportation & logistics: Comparing sample average approximation with simheuristics. In *Proceed. 2019 winter simulation conference, Maryland, USA. December 8-11.*

Lourenço, H. R., Martin, O. C., & Stützle, T. (2010). Iterated local search: Framework and applications. In M. Gendreau, & J. Y. Potvin (Eds.), *Handbook of metaheuristics* (2nd ed., pp. 363–397). New York: Springer.

Melo, S., Macedo, J., & Baptista, P. (2017). Guiding cities to pursue a smart mobility paradigm: An example from vehicle routing guidance and its traffic and operational effects. *Research in Transportation Economics*, *65*, 24–33.

Onggo, B. S., Panadero, J., Corlu, C. G., & Juan, A. A. (2019). Agri-food supply chains with stochastic demands: A multi-period inventory routing problem with perishable products. *Simulation Modelling Practice and Theory*, *97*, 101970.

Panadero, J., Currie, C., Juan, A., & Bayliss, C. (2020). Maximizing reward from a team of surveillance drones under uncertainty conditions: A simheuristic approach. *European Journal of Industrial Engineering*, *14*(4), 485–516.

Pillac, V., Gendreau, M., Gueret, C., & Medaglia, A. (2013). A review of dynamic vehicle routing problems. *European Journal of Operational Research*, *225*(1), 1–11.

Psaraftis, H. N., Wen, N., & Kontovas, C. A. (2016). Dynamic vehicle routing problems: Three decades and counting. *Networks*, *67*(1), 3–31.

Reuter, C., & Spielhofer, T. (2017). Towards social resilience: A quantitative and qualitative survey on citizens' perception of social media in emergencies in Europe. *Technological Forecasting and Social Change*, *121*, 168–180.

Taniguchi, E. (2014). Concepts of city logistics for sustainable and livable cities. *Procedia - Social and Behavioral Sciences*, *2*(151), 310–351.

CHAPTER TEN

Blockchain in a Smart City: Its applications and a selection framework[☆]

10.1 Introduction

In this chapter, one of the emerging technologies, namely blockchain, is discussed to create the foundation of technology and information management in the peer-to-peer network to enhance the functionalities and reliability of connected systems in smart cities. Three areas are involved in the blockchain discussion, namely (i) definition and overview of blockchain, (ii) features of blockchain, (iii) applications of blockchain in daily life. A case study in blockchain technology selection was also included.

The development of a smart city aims to provide a better quality of life to the citizens by optimizing the efficiency of city operations and services through various channels of data collection, like citizens, smart devices, and buildings. Nonetheless, the smart city concept is still developing, and the problems of data privacy and system security have been raised in recent years. Blockchain technology can foster the development of a smart city as it enables highly reliable and transparent data exchange in a decentralized nature. A recently proposed three layers of the blockchain interoperability model can reference applying blockchain in the smart city. The three-layer model for blockchain deployment includes the business model, platform, and infrastructure layers. Each layer contains several aspects, governance, data standardization, commerce, the legal framework in the business model layer; consensus mechanism, smart contract, authentication and authorization in the platform layer; hybrid cloud, managed blockchain, proprietary components in the infrastructure layer (World Economic Forum, 2020).

☆ With the collaboration of Dr Paul Tsang, Project Collaborator, The Hong Kong Polytechnic University, Hong Kong SAR.

Collective Intelligence for Smart Cities
https://doi.org/10.1016/B978-0-12-820139-8.00006-1

This framework briefly suggests a blockchain solution that can be applied to the development of a smart city. The case study also throws light on the background information and industrial criteria in various smart communities. Then it surveys the possible use of blockchain in a healthcare application in smart cities.

10.2 Definition and overview of blockchain

The concept of blockchain was firstly initiated by Satoshi Nakamoto in his white paper, titled "Bitcoin: A Peer-To-Peer Electronic Cash System" in 2008 (Nakamoto, 2008). His paper mainly described the flow of online transactions for electronic cash. A peer-to-peer (P2P) network between various parties can be completed without involving any third-party financial institutions. Blockchain was the key technology to achieve trust and consensus in the computer network. In the traditional public ledger and electronic cash system, the centralized financial institutions have an intermediate role in managing and balancing the cash flow and transaction activities. However, most of the services mentioned above require handling charges paid by users, and the whole process is not transparent and easy for traceability. The cryptocurrency was established to provide secure transactions using cryptography to tackle the challenges, while Bitcoin was the first application of the above concept.

Furthermore, Zheng et al. (2018) summarized that blockchain was constructed in a sequence of blocks, which keeps a complete list of activity/transaction records similar to the traditional public ledger. Specifically, the beginning block in the blockchain is called the genesis block, and the subsequent blocks are added one by one after the genesis block. Nofer et al. (2017) defined blockchain technology as a disruptive technology that influenced and triggered the changes in various industries, handling online transactions and intermediation and ownership verification in the modern business environment. Pilkington (2016) mentioned that the creation of blockchain could eliminate the double-spend problem by adopting asymmetric encryption in blockchain technology. Every node in the blockchain network was aware of the activities and transactions. Once the node is validated, the information and data blocks are publicly announced and synchronized to all the nodes in the P2P network. Therefore, the presence of blockchain technology creates mutual trust and consensus among the users in the computer network.

Thus, the data managed by blockchain technology becomes reliable and secure. Although blockchain technology is deemed a disruptive technology to most industries, the foundation for most industrial applications is strengthened in terms of reliability, security, and mutual trust. By adopting the concept of decentralization in the P2P network, the business models are modified and revamped to create more business values and facilitate business sustainability. Generally speaking, blockchain applications contain a chain of data blocks distributed in the peer-to-peer (P2P) network. Differing from the traditional client-server approach, the peer nodes in the P2P network can be connected and synchronized with each other. The users themselves can support the work of data creation and validation under a specific consensus mechanism.

Regarding client-server-based applications, the data are managed in a centralized manner, where the users can access and interact with the applications by their own devices and portals. Therefore, there is a host server or party to control all the activities, data acquisition, storage, and transmission to achieve the objectives of the applications. However, blockchain applications in the P2P network adopt the decentralization strategy. Every user in the network has the same set of synchronized data instead of storing everything in a centralized server and database. The visibility and transparency of the data management are thus improved, while the system vulnerability, namely the single point of failure, can be eliminated.

In general, there are three types of blockchains, namely public blockchain, consortium (or federated) blockchain, and private blockchain (Lai & Chuen, 2018; Lin & Liao, 2017). Fundamentally, the public blockchain is open for the public who want to participate in the P2P network. The blocks are created and verified through an agreed consensus mechanism, for example, Ethereum and Bitcoin. Consortium blockchains refer to the partially public distributed network, where some nodes are selected in advance to manage the entire P2P network. The data in the blockchain can be open or restricted, depending on the functions and objectives of the consortium blockchains. Lastly, in private blockchains, all the nodes are restricted for a particular organization or company to achieve its blockchain goals. The organization strictly controls data access and management. Table 10.1 shows the comparison between the above three types of blockchain in data management, vulnerability, efficiency, consensus determination, and data accessibility. The selection of types of blockchain should strike a balance between their advantages and disadvantages.

In the following, the process flow and considerations for blockchain deployment are discussed to establish a big picture of blockchain. In other

Table 10.1 Comparison between the public, consortium, and private blockchains.

	Public blockchain	**Consortium blockchain**	**Private blockchain**
Data management	Decentralized	Partially decentralized	Centralized
Vulnerability	Low	Medium	Medium
Efficiency	Low	High	High
Consensus determination	All nodes	Selected nodes	Privately controlled
Data accessibility	Fully open	Partially open	Restricted

words, the journey of block creation and verification is illustrated. In the blockchain, the data block structure contains the data of index, timestamp, data, self-hash value, previous hash value, and nonce, as shown in Fig. 10.1. The index shows the sequential order of the data block, which indicates the particular blocks' position. Timestamp and data refer to the time of data creation and the data to be stored in the blockchain, such as transaction activities. The self-hash value is the encryption result of the blocks by using a hash algorithm, such as SHA-256. The hash algorithm is the one-way function to convert the data and message to a string and value, which is infeasible to be reverted. Hence, a brute force attack is the only solution to solve the hash values recently. The entire encrypted string can be totally different when there is a little change on the data by using the hash algorithm. For instance, SHA-256, which is under the family of SHA-2, generates the hash value with the length of 256 bits by inputting the specific messages and data. Previous hash is referred to the hash value in the previous block, i.e., the block n-1. Also, a nonce is a random or pseudo-random value to modify the block messages to fulfill the particular target difficulty in block validation.

To create the blockchain, the first block at the beginning of the whole blockchain is called the genesis block, which has the "0" value in the index

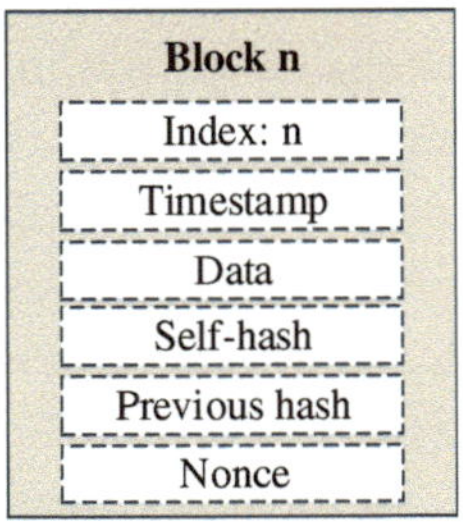

Fig. 10.1 Generic block structure in the blockchain.

and previous hash value. To create the data block in the blockchain, asymmetric cryptography is applied to secure the creation process. Asymmetric cryptography is the improved version of symmetric cryptography, where the keypairs become a private key and a public key instead of using a shared key (Kravitz & Cooper, 2017; Niranjanamurthy et al., 2018). The private key has the role of a node's password and an account with the identification information, but it is not publicly available and the owners should not share it with others. The functions of a private key are to access, authorize, and control the accounts for the activities in the network. The public key is similar to a username, which can be shared with anyone. There is an association between the public key and private key, but it is infeasible to derive the private key from the public key. For example, for sending confidential data from sender to receiver, the data can be encrypted by the receiver's public key, and the receiver can decrypt by using its private key. Subsequently, digital signatures are proposed by using asymmetric cryptography in the blockchain as a promising tool for sending messages and data between various parties. For instance, if user A would like to distribute a particular message in the distributed network, the message can be encrypted using A's private key. Other users can decrypt it by using A's public key. Thus, the message is digitally signed by user A. Also, if user A would like to send another message to user B privately, the message should be encrypted by using A's private key and B's public key accordingly. The message can only be opened by using A's public key and B's private key. Hence, the message can only be received by user B. In particular to cryptocurrency applications, the cryptocurrency address is used to represent the public key. It is derived from the public key utilizing the one-way hash algorithms. Overall, the nodes can effectively create the messages, data, and transaction activities in the P2P network.

The entire blockchain is the series of blocks created and validated in the distributed P2P network, as shown in Fig. 10.2. Strictly speaking, each block

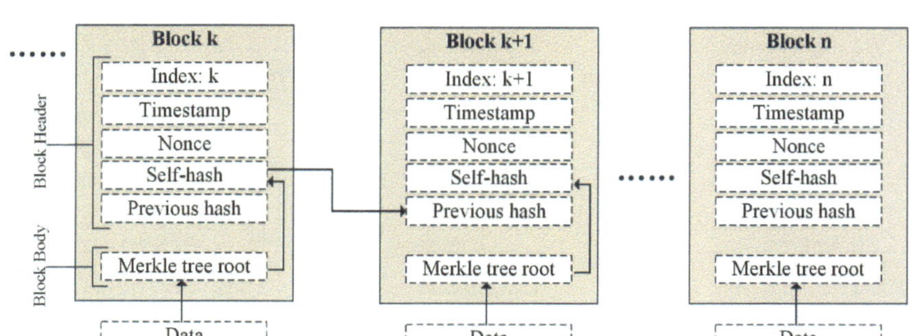

Fig. 10.2 A sequence of created and validated blockchain.

can be defined to have (i) block header and (ii) block body. The block header contains the information about the block itself, including the index, previous hash value, timestamp, difficulty target, and nonce. The difficulty target refers to the restriction on the results of the hash value by entering the messages and data through the adjustments of the nonce. It is used to control the synchronization time and interval of the entire blockchain application. In the block body, it is mainly for the messages and data to be stored.

However, when there are more than one messages and data to be encrypted, the Merkle tree is applied to formulate the Merkle tree root in the block instead of putting all the hash values derived by the data in a single block. Merkle tree is the binary tree to manage multiple hash values from the transactions, data, and messages (Yuan & Wang, 2016). Fig. 10.3 shows the example of creating a Merkle tree by six corresponding data, i.e., A, B, C, D, E, and F. The data can be converted to the corresponding hash values by adopting the hash function. The hash values are then combined two by two, for example, $H_A + H_B = H_{AB}$. Eventually, the entire Merkle tree can be established to manage multiple data in the blockchain effectively, and the top of the Merkle tree is called the Merkle tree root. If the odd number of data is considered, the last data is duplicated (with itself) to complete the calculation of the Merkle tree root.

The consensus algorithms are essential to address various industrial requirements and situations to achieve mutual trust and consensus in the blockchain. In this section, five common consensus algorithms which are actively operating are discussed, namely Proof of Work (PoW), Proof of Stake (PoS), Proof of Burn (PoB), Proof of Importance (PoI), and Proof of Weight (PoWeight).

(i) Proof of work (PoW)

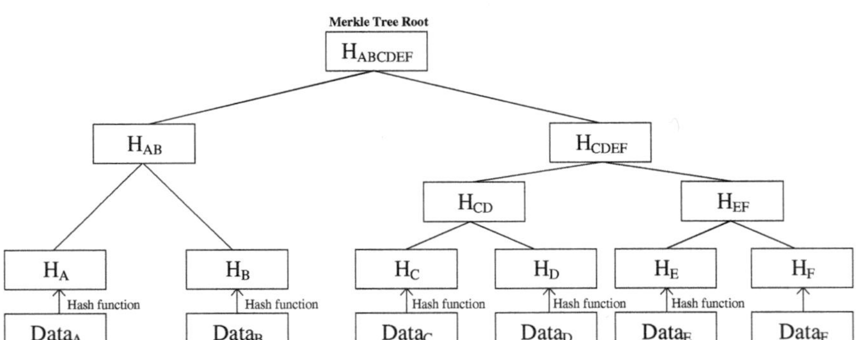

Fig. 10.3 Structure of Merkle tree in blockchain.

PoW is the most fundamental and traditional consensus algorithm to achieve consensus in public blockchains (Pilkington, 2016). To validate the created blocks, all the nodes in the public blockchain compete with each other to solve the cryptographic puzzle, namely the calculation of the hash value with fulfilling the specific target difficulty. Once the solution is found by using a brute force attack, the winning node will be broadcasted over the network, and the other nodes are used to verify the results. This method is regarded as a secure and reliable blockchain application due to the properties of a one-way hash algorithm and difficulty target. The difficulty target refers to how many "0" s are set at the beginning of the output hash values. In the PoW, the entire process is called mining, and the miners attempt to be the first problem solver. To mine the entire blockchain, a high level of computational resources and power in the whole P2P network is required, which may create a counter-effect to the sustainability of the blockchain applications. The energy wastage from the nonwinning nodes cannot be avoided. Suppose some malicious nodes and miners attempt to alter the data blocks in the blockchain. In that case, they must control more than 50% of the computational power and resources in the P2P network, which is almost impossible. The above malicious activity is named a 51% attack in the blockchain. Therefore, the PoW is a good beginning to blockchain development and applications. There is still room to modify and improve the consensus algorithms concerning their effectiveness, efficiency, and sustainability.

(ii) Proof of stake (PoS)

PoS is fundamentally different from PoW, but it is also developed to achieve the consensus between blockchain users. There is no concept of mining and miners in the PoS, but the block forging and validators. For the block forging process, the validators are selected according to their wealth or economic stake in the network. The entire consensus mechanism becomes the probabilistic approach instead of the deterministic approach (Kiayias et al., 2017). In the blockchain network, by using PoS, it is infeasible to control and tamper the blockchain as the validators are challenging to have a 51% or more economic stake in the entire network. In addition, coin age is another factor together with an economic stake. The coin age is defined by (the product of) the number of coins and the time of holding the coin in the wallet. In other words, the validators who hold a large amount of coin for a long time can be selected with a higher likelihood. Since PoS does not require competition between the validators, the energy consumption for operating consensus algorithms is significantly less than PoW, resulting in better system sustainability.

(iii) Proof of burn (PoB)

To concern the sustainability of blockchain, PoB is developed to be the energy-efficient consensus algorithm. The users need to invest their coins in the blockchain applications by burning them (Zheng et al., 2018). In technical senses, the users send a certain amount of coins to the irretrievable address, or called eater address, to obtain the lifetime privilege to mine the blocks on the blockchain. The above activities can be regarded as investments from blockchain users to obtain future benefits. The more the number of coins burnt, the higher the likelihood of them mining the blocks. Afterwards, the rewards can be gained at a later period. The PoB defines that the users are dedicated to the PoB blockchain activities so that the miners can be trustworthy on the blockchain. Similar to the advantages of PoS, the requirements on hardware and electricity costs can be significantly reduced by using the PoB. The selection criterion to mine the blocks is changed from level of economic stake to the willingness in blockchain investment.

(iv) Proof of importance (PoI)

PoI is the mechanism to determine which network users are eligible to harvest a block to the blockchain, where the processing is called block harvesting (Bach et al., 2018). The users with higher importance can have a higher chance to be selected for block harvesting. Differing from PoS, PoI considers three major factors to achieve the consensus, including (i) number of vesting coins, (ii) number of transaction partners, and (iii) number and size of monthly transactions. PoI is developed to solve the challenges of PoW and PoS. On the one hand, users in PoW require powerful computing systems to compete with other users in the network, which may cause excessive energy consumption and harm the environment. In contrast, users in PoS are de-motivated to spend their economic stake as most of them tend to keep the coins for a long period to fight for a greater chance to be validators. The motivation of transactions and activities in PoS is relatively low. PoI has made the balance for the above considerations to create a positive attitude on blockchain applications.

(v) Proof of weight (PoWeight)

PoWeight creates the first step to extend the blockchain to other industrial applications. In PoWeight, the blockchain users are assigned with the corresponding weights equivalent to their contribution to the network (Avasthi & Saxena, 2018). To prevent double-spending attacks on the blockchain, most of the weighted fraction has to belong to the selected and trustworthy users. PoWeight provides a customizable and scalable consensus algorithm to blockchain applications. Compared with PoS, the additional weight factors, such as the amount of interplanetary file system (IPFS) data, can be considered to enrich the coin-age method. Such an approach can also contribute to other blockchain applications and consensus algorithms, which are not limited to the PoS development.

After that, six more consensus algorithms that came up in a later stage are also briefly introduced below. They are usually found in different Blockchain-as-a-Service (BaaS) platforms and frameworks, e.g., Proof of Elapsed Time (PoET), Sumeragi, pBFT, IBFT, Raft and Kafka.

(i) Proof of elapsed time (PoET)

Proof of elapsed time (PoET) is established by Intel and implemented in their blockchain platform Hyperledger Sawtooth. It is developed based on Proof of Work (PoW), which is under the class of Proof-of-X consensus protocols. This algorithm is the improvement of PoW, which removes the computational competition by using the waiting time. In PoET, miners are assigned a waiting time, and they are required to create a new block once the waiting time is ended. Every single miner has an equal chance to gain the authority to mine a new block. All of them are assigned a waiting time randomly. The miner with the shortest wait time will gain the mining responsibility. The mechanism guarantees that the participating miners are selected randomly for the shortest waiting time, not the shortest time selected deliberately by the miners to win. Thus, PoET can eliminate the waste of energy for the computational competition compared with PoW. The participants' miners compete to mine the block based on the target difficulty according to their computing power. The efficiency of PoET performed better than the PoW since the miners can be released and switched to other tasks in the waiting time.

Moreover, PoET provides a secure environment for block publishing since it generates code by a "trusted execution environment" (TEE). TEE safeguards memory by the processor in computing devices. Thus, the code of hardware enforcement can be preserved, and data inside can be confidentiality and integrity protected. Code authorized by the operator or owner, so external participants cannot alter it. Thus, PoET can prevent the malicious participants from faking short wait times and control the chain of block creation under hardware associated with TEE. TEE also can provide externally verifiable attestations and tamper evidence, so the result of PoET can be verifiable by external participants or parties. The transparency of the network can be enhanced.

(ii) Sumeragi

Sumeragi is a consensus algorithm offered by Hyperledger Iroha designed for permission private blockchain. Sumeragi adapts Byzantine Fault Tolerance (BFT) features in tolerating faulty nodes and exploits a global sequence that considers the total network size 2f +1, which includes considering the normal and malicious nodes, where f denotes the number of malicious nodes in the network. More than a half signatures are required to reach a consensus on every transaction to validate transactions, which makes only the set of normal nodes capable of contributing to the consensus procedure. The first set of nodes is used for verifying the transaction. Then, the transaction will pass to another validator set to evaluate the authenticity of signatures and contents.

(iii) pBFT

Practical Byzantine Fault Tolerance (pBFT) is ideal for working efficiently and practically in asynchronous systems. This consensus algorithm is available on Hyperledger Sawtooth. A new data block is determined during each round to classify transactions in the pBFT process. There are three different stages which are preprepare, prepare and commit. Since the pBFT requires a total network size of = 3f + 1 in which f represents the number of fault nodes, the transaction must be verified by at least ⅔ of the total nodes. It can tolerate ⅓ of malicious nodes in the system. For this reason, pBFT ensures the security of the transaction in the permissioned blockchain with known users. Therefore, the block verification and finality process can become efficient.

(iv) IBFT

Istanbul-BFT (IBFT) is a kind of consensus algorithm using byzantine fault tolerance offered by Quorum. It also uses a three-phase consensus, which contains "priori-prepare," "prepare" and "commit." The system can tolerate ⅓ malicious/fault nodes since it requires $N = 3f + 1$ validation, which f represents the fault nodes. IBFT enjoys the advantage of immediate transaction verification and less resource needed in terms of hardware compared to PoW. Also, the immediate block finality can be ensured in this consensus algorithm.

(v) Raft

Raft is an open-source software platform based on Crash Fault Tolerance (CFT) mechanism and is used by Hyperledger Fabric. It is designed to be easily understandable. Raft is decomposed into independent subproblems and can address everything the system needs. The server states in Raft can be a leader, a follower, or a candidate, that the system has to elect a leader to achieve goals. The leader is in charge of the log replication, which accepts requests from clients. The requests should consist of a command executed by the replicated state machines in the cluster. Once the leader accepts the request as a new entry, it will be forwarded to the followers as AppendEntries messages. Once the majority of the followers have replicated the entry, the leader receives a confirmation of that. Then, the leader applies the entry to its local state machine and considers the request as "committed." Followers will also apply the entry to their local state machine once they learn that a log entry has been "committed." This mechanism ensures that the consistency of the logs in servers (through the cluster) and the safety rule of log matching are respected. According to the log matching property, if two logs contain an entry with the same index and term, the logs are identical in all entries up through the given index. The logs will become inconsistent if the leader crashes, some logs will be left and not being adequately replicated through the cluster by the old leader. The new leader has to handle the inconsistency by forcing the followers to duplicate its log. The leader has to compare its log with each of the logs of followers, find out the latest entry they agreed on, delete all entries after that agreed entry and replace it with its log entries. These steps help restore log consistency in a cluster subject to failures.

(vi) Kafka

Kafka is an open-source software platform based on Crash Fault Tolerance (CFT) mechanism developed by the Apache Software Foundation, written in Scala and Java. It is a distributed event streaming platform, which event streaming is a practice of capturing data in real-time from event sources. Kafka is not a blockchain. Kafka allows users to publish and subscribe to any data on any systems or real-time applications, provides decentralized and durable storage that acts as a "source of truth" with support of distributed and immutable log of records, and allows for on-the-fly processing which lets users perform data joins within a stream. Thus, Kafka is capable of many real-world enterprise blockchain projects which allow decoupling applications/clients and role-based access control to data.

Kafka stores key-value messages from producers in a cluster of one or more servers, also known as brokers. The messages include different topics and partitions and are ordered by their offsets within a partition with a timestamp, while consumers can read the messages from partitions. Kafka has five major APIs, including (1) Admin API—to manage and inspect brokers, topics, and other Kafka objects; (2) Consumer API—to read the topics and process streams of records; (3) Kafka Connect API—to execute the reusable data from (2) to external systems that allow them to integrate with Kafka; (4) Kafka Streams API—to convert the input streams to output and produce the result; (5) Producer API—to write the streams of events to one or more topics. Kafka can be considered as a conjunction with blockchain, but it lacks tamper resistance and encrypted payloads.

The above content shows the functions, values, and overview of blockchain technology. However, the necessity of adopting blockchain is also a concern to businesses and organizations. The following summarizes six primary considerations for deciding on blockchain adoption:

a. Do you need a common shared database in your smart city solution?
b. Will you involve multiple parties in the computer network?
c. Is there any conflicting incentives or mistrust between the parties?
d. Are there any differences in the rules for governance between the parties?
e. Do you need an objective and unchangeable log of records?
f. Do you have any standardized rules behind the activities/transactions?

If the results of the above six considerations are positive, the situation can be treated as an appropriate use case for adopting blockchain technologies. Otherwise, the need for blockchain technology should be re-evaluated and investigated. Blockchain technology has been an emerging technology in recent years, but misuse of such a technology should be avoided.

10.3 Features of blockchain

After applying blockchain technology, certain benefits and advantages can be obtained to enhance data management and transaction effectiveness. This section summarizes five main features of using blockchain in real-life situations, including (i) prevention of data tampering, (ii) decentralization of data management, (iii) improvement of system security, (iv) use of smart contracts, and (v) mutual consensus establishment.

(i) Prevention of data tampering

With blockchain technology, there are two basic mechanisms to prevent data tempering and eliminate the vulnerability. First, when mining the blocks or establishing the blockchain, the longest chain rule is applied to identify the valid chain of the data blocks (Kiayias et al., 2017). Since the longest blockchain implies the most effort that has been put to create and validate the blocks, reliability and trustworthiness can be guaranteed. Fig. 10.4 illustrates the longest chain rule in the blockchain. In particular to the PoW mechanism, when someone wants to add block A, the block validation cannot be completed due to the existence of block k + 2 and the inconsistency of the last block. Assuming that block k + 4 is the last block on the blockchain at this moment, the time for creating and validating blocks B, C, and D after block k + 2 should be longer than the time for creating and validating block k + 5. Thus, attacking this longest chain rule in this way is nearly impossible in the PoW environment. Simultaneously, the normal block creation and validation process are continued, such that the malicious nodes are required to outperform more than half of the peer nodes. Unless the users can control at least 51% of the computational resources, which is difficult to achieve. The above idea can also be extended to other consensus mechanisms, such as PoS and PoI. Tampering the data on the blockchain can be achieved if only a user can dominate and control at least 51% of the share and stake in the entire P2P network. However, it is challenging to meet the above criterion so that the validated data on the blockchain can be securely protected.

(ii) Decentralization of data management

As mentioned above, the blockchain is operated in a P2P network, which implies that the blockchain's data is decentralized. There are

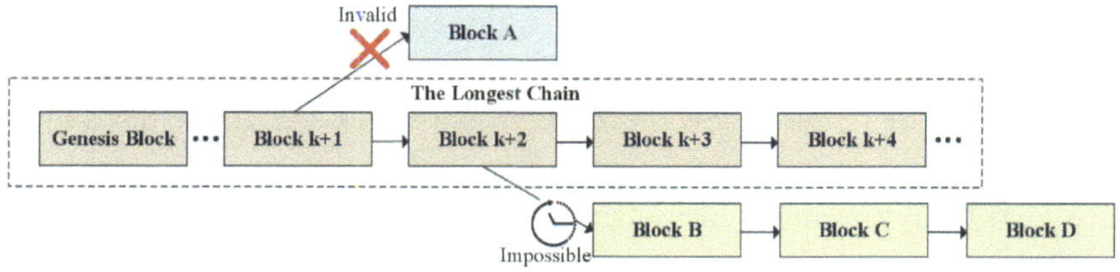

Fig. 10.4 Illustration of the longest chain rule on the blockchain.

three dimensions to explain blockchain's characteristics of decentralization and to evaluate a blockchain. First, blockchains are politically decentralized for users because there is no anybody or a central authority to control the network. Secondly, blockchains are architecturally decentralized without the central infrastructural point of failure. This characteristic also improves the security of the blockchains. Even one user is being attacked, other nodes in the blockchain network can recover the data to other affected users. Thirdly, blockchains are operated in a logically centralized manner because there is a commonly agreed consensus mechanism and system operations for the blockchains. The behavior of the entire blockchain system acts as a centralized system to reach the objectives and goals of the system. An interplanetary file system (IPFS) is proposed to achieve the above decentralization in practice. IPFS addresses the challenges of the existing Hypertext Transfer Protocol (HTTP), which is a request-response method, such as a tremendous amount of request-response activities and broken links of communication (Chen et al., 2017). Fundamentally, the access by using HTTP is to ask the particular network location, while the access by using IPFS is to ask the certain files in the network. With the integration with the world wide web (WWW), IPFS tends to overcome the future challenges for the coming era of 5G and artificial intelligence, including effective management of petabyte datasets, the spread of real-time media streams, and data security and authentication.

(iii) Improvement of system security

One of the core features of using blockchain is the immutability once the blocks are created and validated. The data on a blockchain is almost impossible to be altered. Therefore, most organizations and enterprises treat the blockchain as a hot technology and improve their system security by integrating enterprise systems with a blockchain. Currently, businesses and enterprises are spending tens of thousands of resources to maintain cyber security and prevent hacking activities. The above-mentioned decentralized data storage is to improve security. It is no longer practical to attack a single point of node or server unless the hackers can attack the entire distributed network simultaneously. Secondly, under the internet of things (IoT) environment, the edge devices are always the weaknesses to be attacked so that the use of blockchain can strengthen its autonomy for security decisions. A consensus can be established between the authorized devices and IoT nodes so that other nodes which behave suspiciously can be blocked. In addition, blockchain also supports the near real-time data transmissions and regular synchronization among the devices. Thirdly, blockchain technology can also create a safer environment for the domain name server (DNS). Traditionally, DNS is centralized and vulnerable to breaking the connection between website names and IP addresses. Also, the DNS can be paired with a (distributed denial of service) DDoS attack to terminate the web servers and services. When having blockchain for DNS, the domain information can be managed immutably on the distributed ledger. The connections between website names and IP addresses can be made by using smart contracts.

(iv) Use of smart contracts

Under the blockchain environment, smart contracts are developed to exchange value between parties in a transparent and conflict-free manner while acting as the machine-to-machine mechanism (Christidis & Devetsikiotis, 2016). Smart contracts are prepared as programming codes that are publicly available on the distributed ledger. When the smart contracts are activated by meeting specific criteria, the details and statements in the contracts will be executed, and the obligations will also be automatically enforced. The middleman and relevant services to maintain the legal obligations can be eliminated by using smart contracts. Overall, the smart contracts on the blockchain have five significant advantages, including autonomy, trust, automatic backup, cost-effectiveness, and accuracy. It can genuinely guarantee the specific outcomes and obey the conditions, and litigation can never occur in the blockchain environment.

(v) Mutual consensus establishment

The consensus is one of the most significant breakthroughs in blockchain technology. Although data security and management can become more effective and accurate, establishing trust between various parties is the core value for blockchain adoption. By using agreed consensus algorithms and mechanisms, the users are willing to trust the data displayed on the blockchain, which is different from other centralized approaches. Establishing a mutual trust can strengthen most business and industrial applications, such as supply chain management and cryptocurrencies. In real-life situations, the trust between the strange and unknown users is difficult and time-consuming to be formulated. Due to the above features and characteristics of blockchain technology, the information on the blockchain is objective, accurate, and equivalent to every user. Apart from the cryptocurrencies and initial coin offering (ICO) businesses, the blockchain can be further applied to other industries by creating trust and improving the reliability of the businesses, resulting in better sustainability and scalability.

10.4 Applications of blockchain in daily life

This section discusses blockchain applications in daily life activities. It is beyond cryptocurrencies to investigate their scalability and adaptability. Because of that, supply chain traceability and healthcare are two areas that are going to be illustrated.

(i) Supply chain traceability

Supply chain traceability has always been an active research area related to our daily life to trace the products from the origin to the point of consumption across the complex supply chain network. The success of the supply chain traceability needs to provide transparent, reliable, and tamper-proof information regarding its credential data, logistics activities, and product information. In particular, to the perishable products, the role of supply chain traceability is of the utmost importance. The product description provided by the suppliers or vendors was highly uncertain. For example, the suppliers or vendors declare that the products were planted in an Australian farm, but they were not. The consumers cannot verify the truth because the suppliers provide all the product information, and the data stored in a centralized manner can be tampered with (or with fake data). After customers purchase the products online, the suppliers will provide the logistics information for their customers to track the products, yet, no one can ensure the accuracy of the related information. The information is provided by the logistics services providers so that the information is dominated by one party, and this party can write whatever they want the customers to know. Under this circumstance, customers may receive misleading or incomplete information that threatens their right to know. Therefore, blockchain technology should be applied to strengthen the values of supply chain traceability for meeting the needs for handling perishable products and the environment of e-commerce businesses. Lu and Xu (2017) proposed an adaptive blockchain-based system for product traceability, called OriginChain, to connect suppliers/vendors, traceability service providers, and lab testing companies. In the blockchain, two types of data are stored, i.e., hash of traceability certificate or photos, and a small amount of traceability information confronted to the regulations.

The raw files regarding the traceability documents and information are stored in a centralized manner to reduce the capability and burden of blocks. In addition, the decision of putting the data between on-chain and off-chain should be considered according to the requirements of privacy and blockchain performance. Putting more data on the blockchain definitely can improve the security and privacy concerns, but the performance of mining the blocks can become time-consuming and challenging. Figorilli et al. (2018) implemented the blockchain for the traceability of wood along the whole supply chain. The traceability process covers timber marking, cutting, stacking, transport, sawmill processing, production, and selling to final customers. The data acquisition regarding the traceability information is collected by using the IoT technologies, including radio frequency identification (RFID), near field communication (NFC), and barcode. They used the remote database and Azure blockchain workbench to achieve the blockchain implementation for wood traceability.

Although the key ideas of blockchain are simple, its implementation proposes a considerable number of challenges. Theoretically, blockchain technology can achieve the advantages above when employed in the supply chain industry; practically, some difficulties are implementing the industry's blockchain mechanism. Implementing the blockchain into the supply chain has three major difficulties for adopting the technology practically, including (i) lack of system storage capacity and (ii) inefficient and ineffective mining process. Storage capacity has been a significant problem in blockchain because the chain is continually growing. It affects the size of the blockchain and indirectly influences the processing speed of the mining and synchronization. According to Reyna et al. (2018), it is mentioned that only full nodes are required to store the complete chain. However, the storage requirements remain significant. A full node means a node that can entirely validate transactions and blocks. The increased size in blockchain also takes a longer time (than before) to synchronize for the entire P2P network. Secondly, the functionality of blockchain includes routing, storage, wallet services, and mining. All of the functionality is provided by network peers to operate and support the blockchain. As Makhdoom et al. (2018) defined, the function of mining in the blockchain is adding validated transaction information to a block and then broadcasting that specific block onto the blockchain network, which is to be identified by all the nodes. The mining work needs to be finished by miner nodes, and selecting a node to mine a new block is achieved depending on specific lottery schemes. The existing consensus mechanisms are mainly designed for cryptocurrencies, which do not meet product traceability requirements. Therefore, the consensus mechanisms need to be modified and improved to enhance practicality and adaptability.

(ii) Healthcare informatics

Healthcare informatics and data management can be enhanced accordingly since blockchain technology provides a reliable and secure data management protocol. Yue et al. (2016) indicated that the effective sharing of healthcare data was essential to improve the quality of healthcare services. They proposed an application called healthcare data gateway (HDG) to facilitate the data transmission between the stakeholders of healthcare services. Recently, patients' healthcare data have been managed in electronic medical records (EMR) and electronic health records (EHR), central to controlling data sharing and transmission. The blockchain was integrated into the existing ways to improve security and privacy concerns. As shown in Fig. 10.5, three layers, namely secure data storage layer, data management layer, and data usage layer, were presented to organize the adoption of blockchain technology for managing healthcare data. By integrating the blockchain in healthcare data management, the patients do not need to trust third-party companies to handle and manage their personal and private data. The decentralization nature can distribute the data to authorized institutions, such as government and healthcare service providers. The middleman in managing healthcare data can be eliminated, such that the risks of data leakage and tampering can be minimized, while its overhead cost can also be reduced. Consequently, patient privacy can be ensured through the use of blockchain technology.

Although the blockchain has started changing the ecosystem of health informatics, there are several concerns and considerations on the technology adoption. Gordon and Catalini (2018) suggested four dimensions about the challenges of blockchain-enabled patient-driven interoperability, namely (i) transaction volume of clinical data, (ii) privacy and security, (iii) patient engagement, and (iv) incentives. Similar to the applications for supply chain traceability, the design of the blockchain is not for handling

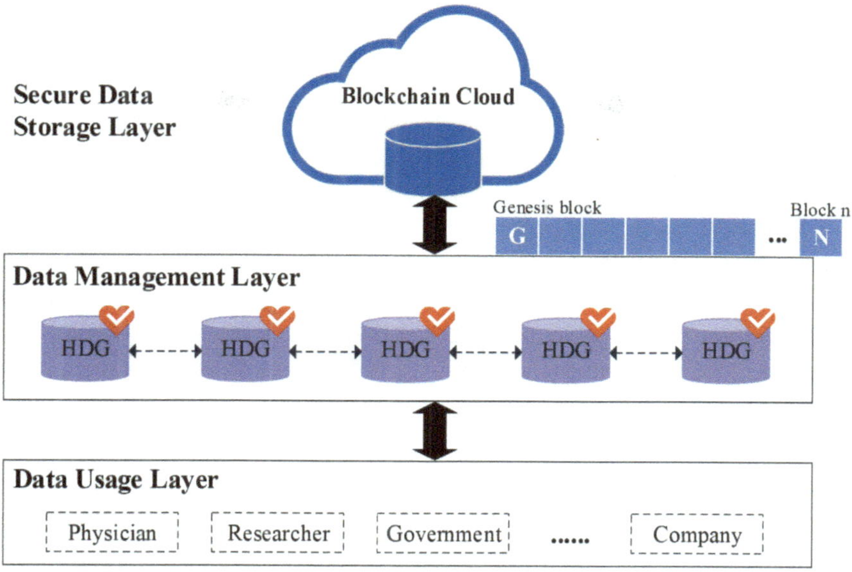

Fig. 10.5 HDG-oriented healthcare eco-system.

and storing bulky datasets. When tremendous data are inputted into the blockchain, the performance decreases significantly, while the time for mining blocks increases significantly. Therefore, the hybrid uses of cloud databases and blockchain should be considered to strike a balance between the performance and efficiency of the blockchain.

Regarding healthcare information, the types of blockchain are restricted to the consortium and private blockchains, where the nodes are permissioned and authorized to create and validate the data blocks. The private and personal information can only be exchanged to the authorized parties instead of the general public. Moreover, patient engagement is essential when deploying blockchain-enabled healthcare solutions, as the patients need to keep their private key and public key to control their accounts. Training and supportive measures, such as bio-authentication technologies, should be provided to assist the adoption of blockchain in healthcare informatics. Last but not least, the consensus mechanisms and algorithms should also be customized to the healthcare industry. Only relying on PoW may cause high costs in energy and equipment maintenance. Therefore, customized algorithms and mechanisms are needed to effectively represent the weight and importance of the stakeholders and patients by considering the potential value gained after using the blockchain solution.

10.5 Different industrial criteria

The use of blockchain is extensively diversified in different industries and can help those industries eliminate their business problems. Four different smart applications in different industries were reviewed in this section, including healthcare management, supply chain management (for fresh meat and automotive), healthcare, and retail. The information gathered in this section may help reveal the features of blockchain, and the important criteria for using blockchain technology in smart city development are explored accordingly. Consequently, various blockchain solutions selection framework is made to support the future blockchain deployment in the industry.

10.5.1 Electronic health record

An electronic health record (EHR) stores patients' information digitally and electronically. This system stores personal information such as age, gender, and medical records. It can share patients' information in public organizations and private clinics to enhance the information flow in healthcare services. Those records will be shared through the internet, enterprise-wide information systems, and other information systems. Sharing health records through the EHR brings benefits to both patients and medical institutions. First, it can improve the quality of treatment since doctors and medical service providers can comprehensively understand the patients' health status. It helps the medical institution identify the high-risk patients and prevent the chance of hospitalization by using data analysis. Besides, it can store the data accurately and securely. Using paper medical records is hard to track patients' health information and time-consuming. It also may be lost, and the data may be outdated. EHR can solve the problem of outdated data and can easily capture more updatable patients' information. Since the patient's file will be modified in each treatment, the risk of the loss of medical incidents can be reduced as the intuitions can gain more accurate information and enhance the effectiveness of the treatment.

However, although there are many benefits in using EHR to record patients' records, some criteria improve EHR. Regarding data privacy, through EHR, around 173 million data have been breached since 2009. Information stored in EHR becomes the target of cyberattacks. EHR brings negative consequences in information processing. Also, in cyberattacks, More and more cyberattacks are launched to the EHR system, affecting at least 200 institutions in the United States. Therefore, there is a need to further enhance the data and system security for the EHR to build a reliable and secure healthcare information exchange worldwide. There are many concerns for facilitating the selection of blockchain settings in the healthcare industry.

10.5.1.1 Concern 1: Security—Information confidentiality

The patient's medical records stored on the blockchain should be secured from any third-party access to ensure the confidentiality of the patients' records. The patient's data include essential information about the patient, such as their medical history, blood group, lab results, X-rays reports, MRI results, and many other related results and reports. All of this information is critical not only to patients but also to the hospital. The records being stored and accessed in the system are only accessible by trusted parties. The system should deny any untrusted third party trying to access the system. With the information being kept confidential from third-party access, the framework would ensure that it would have privacy as well.

10.5.1.2 Concern 2: Security—Data integrity

It measures the trust of the system and information stored in a tamper-proof and reliable way. Using blockchain can ensure information stores are complete and it will not change by unauthorized users. Based on the immutability of the blockchain in nature, the data managed in the blockchain are difficult to be altered, which provides a great extent of trust to end-users.

10.5.1.3 Concern 3: Usability—Accessibility

Using the Role-based access mechanism ensures that every entity of the system is assigned a role. Any third party who is not authorized to have access to the system would not allow accessing the system. Only certain users can view, add, and change the medical record, for instance, doctors and nurses. Authentication allows who can access the information. Blockchain technology is secure and uses specific protocols and mechanisms to keep itself secured from third-party intrusions.

10.5.1.4 Concern 4: Usability—Operability

Operability refers to how user-friendly the software systems are and how well it works. The system should be reliable and work well for both the end-users and the operation teams. Application restarts, server reboots, and load-balancing hacks are the situations that want to eliminate applications from the operation teams. Software systems with well-working help are easier to operate and maintain. Blockchain technology can perform with good operability. First, blockchain is fully organized and has fewer failures. It does not depend on human calculations as it is highly fault tolerant. Accidental failures on the blockchain are not usual. Second, it is less likely to break down as decentralized can prevent malicious attacks since attacking this system is more expensive and time-consuming for hackers. So, blockchain can perform with good operability in the application, which helps the operations to work well in the EHR system.

10.5.1.5 Concern 5: Performance efficiency—Lead time for blockchain synchronization

Time behavior means the capability of the software system to perform its function with appropriate response and processing time and throughput rates. The transaction speed used in the healthcare industry should be as fast as possible since transactions occur anytime and anywhere. Since patients may be sick at any time, the healthcare institution should update the medical record as soon as possible. The blockchain should create a block to record the transaction as fast as possible.

10.5.1.6 Concern 6: Reliability—Availability

The software is used to ensure that systems are running and available most of the time. A blockchain system is needed to run 24 h for the healthcare institution as a transaction occurs anytime, anywhere. Also, a high level of quality of service (QoS) should be guaranteed in blockchain-enabled healthcare information systems. In return, the block creation and validation mechanism should always be available during the system operations.

10.5.2 Retail in luxury goods

Luxury goods are often unique, rare, and precious. High-end fashion, jewelry, and sports cars are examples of luxury goods. Luxury goods are used to represent one's wealth and social status. Thus, luxury brands focus on their uniqueness, image, and reputation. However, as the price of luxury goods is exceptionally high while people are fascinated with them, it arouses the problem of counterfeit goods, especially in the high-end fashion industry. Counterfeit goods ruin the uniqueness and rareness of luxury brands and cause tremendous loss. Counterfeited fashion earns around \$1.2 trillion per year and makes the high-end fashion brand lose approximately \$10.8 billion per year. Thus, the problem of counterfeit goods is adversely affecting the high-end fashion brand. In 2019, a luxury company LVMH cooperated with Microsoft and announced a blockchain platform—AURA, which aims at luxury goods authentication to prevent counterfeit products. Blockchain technology may be a good solution to fix the problem.

10.5.2.1 Concern 1: Products—Guarantees and integrity

The Internet of Things (IoT) means physical objects with a network to connect and exchange data with other devices and systems over the internet. Creating a distributed ledger combined with IoT and blockchain can help prevent counterfeit goods by achieving a robust traceability and product authentication mechanism. Using blockchain technology in luxury retail can store all information in a tamper-proof and reliable way. It can ensure that all information in each stage is correct, which cannot be altered by any other parties. For instance, the suppliers enter the raw materials information, the producers enter the production information, and the sales enter the sales information. Information will be stored accurately and in a tamper-proof way.

Furthermore, the distributor they create can record the data in each transaction stage, like the origin of raw material, producing detail, and its sale location. All goods data will become transparent and authenticated so all stakeholders can track and share transaction data.

Smart tags are unreplicable and allow all stakeholders to capture the complete genealogy of goods. For example, even a fake tag is produced. Still, the tag's data does not align with the authentication mechanism, for example, tag-based, location-based, and object-specific authentication, in the blockchain, so it can instantly know the product is a counterfeit. Once a transaction occurs and is recorded in the blockchain, the transaction data is unchangeable and immutable. Even if a smart tag is duplicated, data in the blockchain will indicate the duplicate products as fake goods since the blockchain can show the product information and the point of sales of the product. Besides, using blockchain with IoT can prevent fake parts of products. The company can use blockchain technology with IoT and can confirm the authenticity of the parts. Thus, it can provide an identification of authentic goods and prevent the loss of counterfeit goods.

10.5.2.2 Concern 2: Products—Stability

Using blockchain technology with IoT can help the company to produce stable and high-quality products. Blockchain with IoT can help inspect the quality of raw material, semiproducts, and finished goods. Since the information in the blockchain cannot be tampered with, if the raw material is not aligned with the information (in the blockchain), it can effectively identify the inferior materials. It can be used in the production stage as well to ensure the quality of finished products. The company can use blockchain technology in the production stage to compare the products according to the description in blockchain and find out the unqualified products. The quality in all production processes can therefore be ensured.

10.5.2.3 Concern 3: Usability—Accessibility

Using a role-based access mechanism in blockchain can allow only authorized users to access the system such that the user privilege can be under effective control. For example, only the suppliers, producers, and sellers can enter and create information in the consortium or private blockchains. Customers are only allowed to view the information of products, but the capability to create and verify the data blocks. Moreover, specific business logic can be modeled in the smart contracts in the blockchain systems, while the logic can be executed automatically.

10.5.3 Automotive supply chain management

The counterfeit part flowing in the automotive supply chain is one of the major concerns from the public, which may lead to user safety. Almost every part has counterfeited versions in the market, such as airbags, engines, wheels, batteries, and windscreens. According to the European Office of Intellectual Property statistics, it is estimated that EUR2.2 billion is lost on counterfeit vehicle wheels. The safety of the consumer and the goodwill of the automotive manufacturers have been negatively affected accordingly. Due to the inferior quality of counterfeit products, the number of claims for replacing defective parts is skyrocketing. To overcome the fact that counterfeit flows in the automotive market. A blockchain solution called PartChain powered by BMW aims to track the automotive components along the product lifecycle. This motivation encourages the stakeholders in the automotive industry to explore blockchain technology because of the data decentralization for recording the data, leading to increased supply chain integrity.

On the other hand, the suppliers may not want to share their sensitive information with other entities. Using blockchain is an ideal solution to balance the interest and privacy concerns of different stakeholders. By using blockchain to record the automotive components with serial numbers, the stakeholders, including the dealers, suppliers, distributors, and third-party logistics, can view the same data. Hence, the entry of counterfeit parts can be stopped owing to the immutability feature in the blockchain.

10.5.3.1 Concern 1: Security—Authenticity

Announce official specifications of parts announced by vehicle manufacturers. Official or unofficial repair centers may base on the blockchain to purchase genuine products and ensure that end-users pay for genuine products. End-users may also verify the products they bought. In other words, the blockchain may streamline the process of parts' identification, verification, and authentication, while the data stored in the blockchain are immutable. Therefore, the part authenticity can be highly guaranteed.

10.5.3.2 Concern 2: Security—Integrity

Data integrity is essential. No one can pretend to be the official manufacturer and release fake or alter the original data. With the secure data management mechanism in blockchain, data decentralization is free from the risk of a single point of failure in the traceability system. If one of the peer nodes is attacked, the data can be effectively recovered from the P2P network. More importantly, the data decentralized in the P2P network is regularly synchronized, and thus the data discrepancy between the peer nodes can be eliminated.

10.6 A case study in blockchain selection in the context of healthcare

In this case study, the Best-Worst method was used to choose a suitable consensus algorithm for a blockchain application in the healthcare industry, which refers to selecting a specific blockchain framework for the system deployment. Kafka, Raft, Proof of Elapsed Time, IBFT, PBFT, and Sumeragi are the possible consensus algorithms in the case scenario. Fig. 10.6 shows the hierarchical structure of the selection framework for healthcare management. Consequently, the appropriate consensus algorithm can be determined to formulate the blockchain application, distributed database management system, and smart contracts.

10.6.1 The first level on the healthcare industry

On the first level, it measures the most important criteria that affect the healthcare industry when it applies blockchain technology. Security, product, performance efficiency, suppliers, reliability, and usability are the measuring criteria for the healthcare industry.

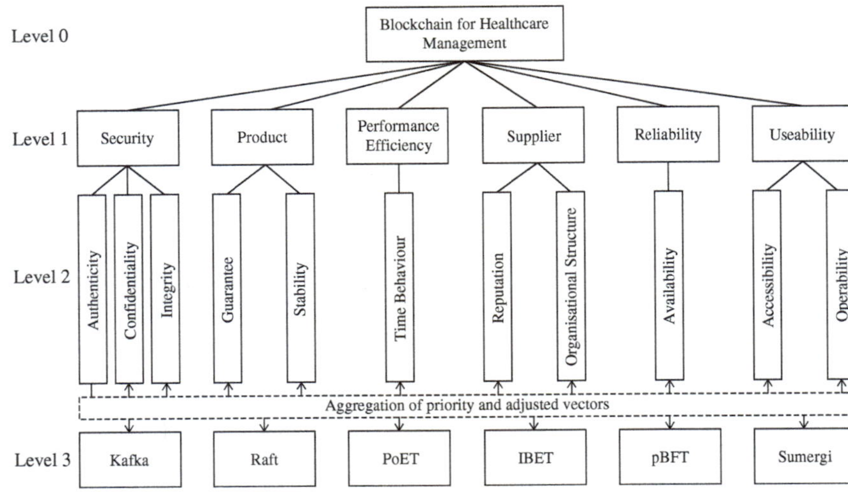

Fig. 10.6 Hierarchy of the selection framework for healthcare management.

Table 10.2 Pairwise comparison of the major criteria in the level 1.

	Criterion 1	Criterion 2	Criterion 3	Criterion 4	Criterion 5	Criterion 6
Criteria name	Security	Product	Performance efficiency	Supplier	Reliability	Useability
Select the best	Security		Select the worst	Supplier		
Best to others	Security	Product	Performance efficiency	Supplier	Reliability	Useability
Security	1	8	4	9	4	3
Worst to others	Security	Product	Performance efficiency	Supplier	Reliability	Useability
Supplier	9	2	5	1	4	6

It is assumed that security is the most crucial criterion for the healthcare industry to consider here. Since protecting customers' private information is essential in the healthcare industry, security is the most important criterion, so the security score is one in best to others as in Table 10.2. Suppliers are the criteria that do not have an important relationship with the healthcare industry. Thus, this criterion will have the lowest mark in best to others which is nine. In Table 10.3, the result of KSI in the first level is 0.076. Using the result of KSI helps to generate the consistency ratio. The calculation of consistency ratio is KSI/consistency index. According to the diagram, we find

Table 10.3 Results on the first level.

Weights	Security	Product	Performance efficiency	Supplier	Reliability	Useability
	0.45191	0.06605	0.13212	0.04171	0.13210	0.17613
Ksi	0.07648		Consistency index	**5.23**		
aBW	**9**		Consistency ration	0.01462		

that the consistency index is 5.23. The result of the consistency ratio is 0.0146. Since the consistency ratio is lower than 0.1, it implies the result is reliable.

10.6.2 The second level of the healthcare industry

The second level measures the subcriteria: authenticity, confidentiality, integrity, guarantees, stability, time behavior, reputation, organizational structure, availability, accessibility, and operability.

In Table 10.4, authenticity, confidentiality, and integrity are the criteria under the Security criteria. The most critical subcriteria is confidentiality since it refers to protecting the customers' private information to avoid data breaches. Since these three criteria have strong relationships related to the health industry, the score range is small in this measurement. As shown in Table 10.5, the result of KSI in the second level is 0.076. According to the diagram, the consistency index is one. The result of the consistency ratio

Table 10.4 Pairwise comparison on the second level under the security.

	Criterion 1	Criterion 2	Criterion 3
Criteria name	Authenticity	Confidentiality	Integrity
Select the best	Confidentiality	Select the Worst	Integrity
Best to others	Authenticity	Confidentiality	Integrity
Confidentiality	2	1	3
Worst to others	Authenticity	Confidentiality	Integrity
Integrity	2	3	1

Table 10.5 Results of using the best-worst method on the second level under the security.

Weights	Authenticity	Confidentiality	Integrity
	0.29167	0.54167	0.16667
Ksi	0.04167	Consistency index	**1**
aBW	**3**	Consistency ratio	0.04167

Table 10.6 Weights of subcriteria in the level 2.

Subcriteria	Weighting
Authenticity	0.291667
Confidentiality	0.541667
Integrity	0.166667
Guarantees	0.5
Stability	0.5
Time behavior	1
Reputation	0.5
Organizational structure	0.5
Availability	1
Accessibility	0.5
Operability	0.5

is 0.0417. Since the consistency ratio is lower than 0.1, it implies the result is reliable.

Since other criteria except security do not contain at least three subcriteria, the weighting of their subcriteria cannot be computed by BWM-solver. Therefore, the weighting of other subcriteria (i.e., guarantees, stability, time behavior, reputation, organizational structure, availability, accessibility, and operability) is additionally designed, as shown in Table 10.6. For criterion that contains two subcriteria, its subcriteria are treated with the same importance. Therefore, the weighting is 0.5 for each one. For criterion that contains only one subcriteria, the weighting is one.

10.6.3 The third level of best-worst method concern to the consensus algorithms

On the third level, it measures which consensus algorithm has the most important relationship with those subcriteria. Authenticity, integrity, guarantees, stability, reputation, organizational structure, and accessibility have the same result in this measurement. The most suitable consensus algorithm

Table 10.7 Pairwise comparison of the consensus algorithm on the level 3.

	Criterion 1	Criterion 2	Criterion 3	Criterion 4	Criterion 5	Criterion 6
Criteria name	Kafka	Raft	PoET	IBFT	pBFT	Sumeragi
Select the best	PoET		Select the worst	Kafka		
Best to others	Kafka	Raft	PoET	IBFT	pBFT	Sumeragi
PoET	7	6	1	5	3	4
Worst to others	Kafka	Raft	PoET	IBFT	pBFT	Sumeragi
Kafka	1	3	7	2	5	4

Table 10.8 Results on the third level.

Weights	Kafka	Raft	PoET	IBFT	pBFT	Sumeragi
	0.05184	0.08813	0.44582	0.10575	0.17626	0.13219
Ksi	0.08294					

for authenticity, integrity, guarantees, stability, reputation, organizational structure and accessibility is PoET as shown in Tables 10.7 and 10.8.

10.6.4 Subcriterion 1: Authenticity

For authenticity (Table 10.9), the most suitable consensus algorithm is PoET. Since the transparency in PoET is in the best performance as the transaction of blockchain can be verified by the external parties, the authenticity can be ensured, and the information in the blockchain will not easily be tamper-proof. IBFT, pBFT, and Sumeragi are the same consensus algorithms (Byzantine Fraud tolerance) class, so they have similar rating scores. Kafka and Raft are with Crash Fraud Tolerance (CFT) consensus algorithm, so they also have similar rating scores. BFT consensus algorithms perform better authenticity than CFT consensus algorithms since CFT consensus

Table 10.9 Pairwise comparisons between (i) authenticity, confidentiality, integrity, guarantees, stability, reputation, organizational structure, and accessibility, and (ii) six consensus algorithms.

	Criterion 1	Criterion 2	Criterion 3	Criterion 4	Criterion 5	Criterion 6
Criteria name	Kafka	Raft	PoET	IBFT	pBFT	Sumeragi
Select the best	PoET		Select the worst	Kafka		
Best to others	Kafka	Raft	PoET	IBFT	pBFT	Sumeragi
PoET	5	4	7	3	2	1
Worst to others	Kafka	Raft	PoET	IBFT	pBFT	Sumeragi
Kafka	2	3	1	5	6	7
Weights	Kafka	Raft	PoET	IBFT	pBFT	Sumeragi
	0.08991	0.11238	0.04407	0.14985	0.22477	0.37902
Ksi	0.07052					

algorithms only prevent system failure. The BFT consensus algorithms prevent not only system failure but also malicious activities. So, the BFT can perform better to safeguard the security of the chain to prevent altered data. pBFT and Sumeragi can perform better in security, so their order will be higher than IBFT. Thus, the order of the BFT consensus algorithm in the best to others will have a higher position than the CFT consensus algorithm.

10.6.5 Subcriterion 2: Confidentiality

For confidentiality, it refers to data confidentiality. BFT consensus algorithms can perform the best in this aspect as it prevents system failure and prevents malicious activities. Sumeragi performs better in malicious activities, as it has two groups of validators into the "A" group and "B" group. "A" group has a leader and proxy tail. If they have byzantine failure nodes, it is better in the "B" group to make the system run. To prevent evil leaders, a node needs to send the transaction to the leader, and the leaders have to submit the receipt by signing or receiving it. If the block is not created at a specific time, the leader will be changed. A signed block will be forwarded to proxy tail. A proxy tail needs to send a commit message at a specific time. If there is a failure of the commit message, a new proxy tail will be selected. So, Sumeragi can double confirm there are no evil nodes to affect the block creation. Thus, the order is one in the best to the others. pBFT can perform better in security than IBFT. Thus, it has a higher position. In crush fault-tolerant, Raft performs better security than Kafka. Thus, it has a higher position. As PoET cannot prevent system failure and malicious activities, it may have a chance of loss of data by hackers or system failure.

10.6.6 Subcriterion 3: Integrity

The reasons for the order are similar to the reasons for authenticity since these two criteria have similar characteristics. Thus, the order of integrity is the same as authenticity.

10.6.7 Subcriterion 4: Guarantees

This criterion is the subcriteria of product which refers to the guarantees of genuine goods. To avoid counterfeit goods, the information of the raw materials, parts, and finished products are correct and authentic. So, the information should not be tampered with or altered by unauthorized parties. PoET consensus algorithms can perform better in this aspect, as the information is transparent in PoET consensus algorithms. The order is one in the best to the other. BFT consensus algorithms have a higher position than CFT consensus algorithms due to data altering.

10.6.8 Subcriterion 5: Stability

This criterion is the subcriteria of the product, which refers to the stability of the quality. To maintain the quality of goods, it needs to confirm all raw materials, parts, and finished products are genuine. The blockchain can help find out the unqualified by comparing the data in the blockchain. Thus, it needs to confirm that all information is correct and authentic. PoET consensus algorithms can perform better in this aspect, so the order is one in the best to the others. BFT consensus algorithms have a higher position than CFT consensus algorithms due to the concern of data altering.

Table 10.10 Pairwise comparison between the time behavior and consensus algorithms.

	Criterion 1	Criterion 2	Criterion 3	Criterion 4	Criterion 5	Criterion 6
Criteria name	Kafka	Raft	PoET	IBFT	pBFT	Sumeragi
Select the best	Raft		Select the worst	PoET		
Best to others	Kafka	Raft	PoET	IBFT	pBFT	Sumeragi
Raft	3	1	7	4	5	6
Worst to others	Kafka	Raft	PoET	IBFT	pBFT	Sumeragi
PoET	6	7	1	5	4	3
Weights	Kafka	Raft	PoET	IBFT	pBFT	Sumeragi
	0.18069	0.43747	0.04755	0.13552	0.10842	0.09034
Ksi	0.10461		Consistency index	**3.73**		
aBW	**7**		Consistency ratio	0.02805		

10.6.9 Subcriterion 6: Time behavior

Table 10.10 shows the result for time behavior. Time behavior refers to the capability of the system product to provide appropriate response and processing time and throughput rates when performing its function under stated conditions. Crush Fault Tolerance performed better in this aspect. Since if there is no system failure, a block will be published. Thus, it can create a block in a short time. Raft has a higher speed of transaction than Kafka. Thus, the order is one in the best to the other. PoET has the worst performance since it needs a waiting time to create a block. The transaction speed performs worst in this aspect.

10.6.10 Subcriterion 7: Reputation

This criterion is the subcriteria of the supplier, which refers to the reputation of suppliers. Choosing a good supplier can enhance the quality of goods. Thus, the reputation of suppliers is the main criterion to select a good supplier. The information of transactions in blockchain helps to select good reputation suppliers. Thus, data transparency is essential in this aspect. Since PoET consensus algorithms have higher data transparency, the order is

Table 10.11 Pairwise comparison between the availability and consensus algorithms.

	Criterion 1	Criterion 2	Criterion 3	Criterion 4	Criterion 5	Criterion 6
Criteria name	Kafka	Raft	PoET	IBFT	pBFT	Sumeragi
Select the best	PoET		Select the worst	Sumeragi		
Best to others	Kafka	Raft	PoET	IBFT	pBFT	Sumeragi
PoET	3	3	1	5	5	6
Worst to others	Kafka	Raft	PoET	IBFT	pBFT	Sumeragi
Sumeragi	5	5	6	3	3	1
Weights	Kafka	Raft	PoET	IBFT	pBFT	Sumeragi
	0.16807	0.16707	0.40966	0.10084	0.10084	0.05252
Ksi	0.09454		Consistency index	3		
aBW	6		Consistency ratio	0.03151		

one of the best. BFT consensus algorithms have a higher position than CFT consensus algorithms due to the concern of data altering.

10.6.11 Subcriterion 8: Organizational structure

This criterion is the subcriteria of the supplier that outlines how certain activities are directed to achieve the goals of an organization. It also determines how information flows within the company. Therefore, timestamps are important in this aspect as it helps track how the goods flow. BFT consensus algorithms have a higher position than CFT consensus algorithms.

10.6.12 Subcriterion 9: Availability

Table 10.11 shows the result for availability. Availability helps keep the software up and running even if a failure occurs. It can eliminate potential points of failure and prepare for the failure of a system by adding redundancy and setting up failover mechanisms. The level of availability can be decided depending on several factors and requirements. Sumeragi has the worst performance in this aspect while PoET has the best performance, which is in the first place.

Table 10.12 Pairwise comparison between the operability and consensus algorithms.

	Criterion 1	Criterion 2	Criterion 3	Criterion 4	Criterion 5	Criterion 6
criteria name	Kafka	Raft	PoET	IBFT	pBFT	Sumeragi
Select the best	Sumeragi		Select the worst	PoET		
Best to others	Kafka	Raft	PoET	IBFT	pBFT	Sumeragi
Sumeragi	5	4	6	3	2	1
Worst to others	Kafka	Raft	PoET	IBFT	pBFT	Sumeragi
PoET	2	3	1	4	5	6
Weights	Kafka	Raft	PoET	IBFT	pBFT	Sumeragi
	0.08851	0.11064	0.05269	0.14752	0.22129	0.37935
Ksi	0.06322		Consistency index	3		
aBW	6		Consistency ratio	0.02107		

10.6.13 Subcriterion 10: Accessibility

Accessibility refers to who can read the data stores depending on the functionality of the industry. For example, an industry like social media platform can let everyone in the world read their data, while some will only let their subscribers have the right to access the contents. PoET performs better in this aspect as it is highly secure. Thus it is placed as the "1st" order. So BFT platforms perform better in the means of security than the CFT, so their order will be higher than the CFT consensus algorithm.

10.6.14 Subcriterion 11: Operability.

Table 10.12 shows the result for operability. Operability refers to the software being user-friendly and how it works well. Since BFT consensus algorithms not only can prevent system failure but also prevent malicious activities. So, it can perform better in this aspect as blockchain in BFT consensus algorithms can work better than other consensus algorithms. Sumeragi will put in order of 1 in the best to others to perform better or prevent the evil nodes. The order of CFT consensus algorithms will follow the BFT consensus algorithms since it can prevent system failure. PoET consensus algorithms perform worst in this aspect as they cannot prevent system failure and malicious activities. The system under PoET consensus algorithms may not perform well.

10.7 Results and limitations

Building a relatively systematic and straightforward way to select a suitable blockchain platform is the advantage of this project. While the disadvantage is that the selected criteria may not cover the specific requirements, the matching method provided an overview for blockchain

Table 10.13 Weights of consensus algorithms over the subcriteria.

	Kafka	Raft	PoET	IBFT	pBFT	Sumeragi
Authenticity	0.05184	0.08813	0.44583	0.10575	0.13219	0.17626
Confidentiality	0.08991	0.11238	0.04407	0.14985	0.22477	0.37902
Integrity	0.05184	0.08813	0.44583	0.10575	0.13219	0.17626
Guarantees	0.05184	0.08813	0.44583	0.10575	0.13219	0.17626
Stability	0.05184	0.08813	0.44583	0.10575	0.13219	0.17626
Time behavior	0.18069	0.43747	0.04755	0.13552	0.10842	0.09035
Reputation	0.05184	0.08813	0.44583	0.10575	0.13219	0.17626
Organizational structure	0.05184	0.08813	0.44583	0.10575	0.13219	0.17626
Availability	0.10084	0.10084	0.05252	0.16807	0.16807	0.40966
Accessibility	0.05184	0.08813	0.44583	0.10575	0.13219	0.17626
Operability	0.08851	0.11064	0.05269	0.14752	0.22129	0.37935

Table 10.14 Measurement of the consistency index of pairwise comparisons.

Pairwise comparison to consensus algorithm	Ksi	aBW	Consistency index	Consistency ratio
Authenticity	0.08294	7	3.73	0.02224
Confidentiality	0.07052	7	3.73	0.01890
Integrity	0.08294	7	3.73	0.02224
Guarantees	0.08294	7	3.73	0.02224
Stability	0.08294	7	3.73	0.02224
Time behavior	0.10461	7	3.73	0.02805
Reputation	0.08294	7	3.73	0.02224
Organizational structure	0.08294	7	3.73	0.02224
Availability	0.09454	6	3	0.03151
Accessibility	0.08294	7	3.73	0.02224
Operability	0.06322	6	3	0.02107

Table 10.15 Composite weights of consensus algorithms.

	Kafka	Raft	PoET	IBFT	pBFT	Sumeragi
Composite rate	0.08788	0.14388	0.20830	0.13239	0.16430	**0.26326**

newcomers to make decisions. When analyzing the above pairwise comparisons, the overall weightings are shown in Table 10.13, while the consistency indexes are calculated for all pairwise comparisons as shown in Table 10.14. The consistency ratios evaluated from the 11 pairwise comparisons are ranged from 0.01890 to 0.03151, which is less than 0.1. It is found that the pairwise comparisons are conducted consistently, and the results from the above analysis are relatively reliable.

After the result aggregation, Table 10.15 shows the composite weights of the consensus algorithms, in which the consensus algorithm "Sumeragi" is ranked as the priority in this analysis. Thus, the healthcare industry could choose a blockchain platform with Sumeragi available, namely Hyperledger Iroha. The use of Hyperledger Iroha to formulate the blockchain-enabled system can comprehensively satisfy the users' concerns and requirements.

The proposed analytics approach on the selection framework is subject to users' perspectives on criteria and consensus algorithms. The findings of this study are not generalized but for a specific scenario. The proposed approach should be re-examined when the scenarios and the group of users are changed. Blockchain development and deployment in the smart city can become more straightforward and effective to overcome real-life problems and pitfalls based on a selected consensus algorithm.

References

Avasthi, A. A., & Saxena, A. (2018). Two hop blockchain model: Resonating between proof of work (PoW) and proof of authority (PoA). *International Journal of Information Systems & Management Science, 1*(1), 128–131.

Bach, L. M., Mihaljevic, B., & Zagar, M. (2018, May). Comparative analysis of blockchain consensus algorithms. In *2018 41st international convention on information and communication technology, electronics and microelectronics (MIPRO)* (pp. 1545–1550). IEEE.

Chen, Y., Li, H., Li, K., & Zhang, J. (2017, December). An improved P2P file system scheme based on IPFS and blockchain. In *2017 IEEE international conference on big data (big data)* (pp. 2652–2657). IEEE.

Christidis, K., & Devetsikiotis, M. (2016). Blockchains and smart contracts for the internet of things. *IEEE Access, 4*, 2292–2303.

Figorilli, S., Antonucci, F., Costa, C., Pallottino, F., Raso, L., Castiglione, M., Pinci, E., Del Vecchio, D., Colle, G., Proto, A. R., Sperandio, G., & Menesatti, P. (2018). A blockchain implementation prototype for the electronic open source traceability of wood along the whole supply chain. *Sensors, 18*, 3133.

Gordon, W. J., & Catalini, C. (2018). Blockchain technology for healthcare: Facilitating the transition to patient-driven interoperability. *Computational and Structural Biotechnology Journal, 16*, 224–230.

Kiayias, A., Russell, A., David, B., & Oliynykov, R. (2017, August). Ouroboros: A provably secure proof-of-stake blockchain protocol. In *Annual international cryptology conference* (pp. 357–388). Cham: Springer.

Kravitz, D. W., & Cooper, J. (2017, June). Securing user identity and transactions symbiotically: IoT meets blockchain. In *2017 Global internet of things summit (GIoTS)* (pp. 1–6). IEEE.

Lai, R., & Chuen, D. L. K. (2018). Blockchain–From public to private. In *Vol. 2. Handbook of blockchain, digital finance, and inclusion* (pp. 145–177). Academic Press.

Lin, I. C., & Liao, T. C. (2017). A survey of blockchain security issues and challenges. *International Journal of Network Security, 19*(5), 653–659.

Lu, Q., & Xu, X. (2017). Adaptable blockchain-based systems: A case study for product traceability. *IEEE Software, 34*(6), 21–27.

Makhdoom, I., Abolhasan, M., Abbas, H., & Ni, W. (2018). Blockchain's adoption in IoT: The challenges, and a way forward. *Journal of Network and Computer Applications, 125*, 251–279.

Nakamoto, S. (2008). *Bitcoin: A peer-to-peer electronic cash system*. White Paper. https://bitcoin.org/bitcoin.pdf.

Niranjanamurthy, M., Nithya, B. N., & Jagannatha, S. (2018). Analysis of blockchain technology: Pros, cons and SWOT. *Cluster Computing, 22*, 14743–14757.

Nofer, M., Gomber, P., Hinz, O., & Schiereck, D. (2017). Blockchain. *Business & Information Systems Engineering, 59*(3), 183–187.

Pilkington, M. (2016). Blockchain technology: Principles and applications. In F. X. Olleros, & M. Zhegu (Eds.), *Research handbook on digital transformations* (pp. 225–253). Edward Elgar Publishing.

Reyna, A., Martín, C., Chen, J., Soler, E., & Díaz, M. (2018). On blockchain and its integration with IoT. Challenges and opportunities. *Future Generation Computer Systems, 88*, 173–190.

World Economic Forum. (2020). *Inclusive deployment of blockchain for supply chains: Part 6—A framework for blockchain interoperability*. White paper https://www3.weforum.org/docs/WEF_A_Framework_for_Blockchain_Interoperability_2020.pdf.

Yuan, Y., & Wang, F. Y. (2016, November). Towards blockchain-based intelligent transportation systems. In *2016 IEEE 19th international conference on intelligent transportation systems (ITSC)* (pp. 2663–2668). IEEE.

Yue, X., Wang, H., Jin, D., Li, M., & Jiang, W. (2016). Healthcare data gateways: Found healthcare intelligence on blockchain with novel privacy risk control. *Journal of Medical Systems, 40*(10), 218.

Zheng, Z., Xie, S., Dai, H. N., Chen, X., & Wang, H. (2018). Blockchain challenges and opportunities: A survey. *International Journal of Web and Grid Services, 14*(4), 352–375.

CHAPTER ELEVEN

Conclusions and future directions of research

Urbanization would be an essential issue in the world. It is predicted that the worldwide urbanization rate would be increased steadily. No matter which country and what aspect are, planning of cities become more and more important owing to rising consideration of citizens' quality of life. Smart city has been widely discussed as a solution to enhancing citizens' quality of life. With the rapid growth of advanced technologies such as IoT, cloud computing, big data analytics, Artificial Intelligence (AI) and blockchain, the numbers of researches and actual applications of smart cities are increasing. For instance, utilizing advanced technologies, several applications such as smart energy networks, smart firefighting, and smart car parking have been discussed in the previous charters to show how such technologies help smart city development.

Firstly, the IoT and cloud, in terms of accessibility, security, data processing, storage, and service sharing, would be intergraded together to enhance the effectiveness and efficiency of different systems. By utilizing these two technologies, various areas in smart cities development would enjoy benefits. There are five layers in the IoT infrastructure:

(1) Business layer on the top, with the IoT applications charge and management.
(2) Application layer, determination of the using applications in the IoT.
(3) Processing layer, handling processes such as information storage.
(4) Transport layer, transmitting and receiving the information.
(5) Perception layer at the bottom, defining the physical meaning of each thing in the IoT system.

Secondly, through data integration, such a large volume of data via IoT can be standardized for data analytics using different techniques such as big data analytics. From the previous chapters, the adoption of big data analytics can analyze massive data with heterogeneous formats such as structured, unstructured, and semistructured data correctly. Thus, big data analytics have been widely applied to various areas such as energy networking,

https://doi.org/10.1016/B978-0-12-820139-8.00013-9

construction, transportation, car parking, and firefighting protection to enhance effectiveness and efficiency in developing a smart city.

Thirdly, cloud computing offers on-demand network access and services of delivery of hardware and software resources in a pay-per-use-basis payment system. The use of cloud computing would provide a solution for data storage in smart cities at a relatively low cost.

However, there are some challenges in adopting advanced technologies, including data privacy issues, design and implementation issues, data accuracy, and legal issue. For instance, errors in data would lower the reliability of the analysis. As IoT contains many types of sensing devices, the accuracy of data collection from people should be concerned. In addition, data security is always the primary concern in a smart city because a huge amount of high-sensitivity data are collected. Therefore, the above challenging areas would be worthy for both academic and industrial practitioners to focus on. This direction can open up more research opportunities to further improve the efficiency, effectiveness, reliability of the smart city development for improving citizens' quality of life in the future.

Acronyms and glossary

3σ The three-sigma value is determined by calculating the standard deviation.
5Vs Volume, velocity, variety, veracity, value
Actuator node A system that converts electrical control signals to physical actions and constitutes the mechanism by which an agent acts on the physical environment.
ADSL Asymmetric digital subscriber line
AHP Analytic hierarchy process
AI Artificial intelligence
AMQP Advanced message queuing protocol
ANN Artificial neural networks
API Application programming interface
BaaS Blockchain-as-a-service
BDCI Big data and computational intelligence
BFT Byzantine fault tolerance
BMW Bayerische Motoren Werke AG
CI Consistency index
CR Consistency ratio
CAN bus A robust vehicle bus standard designed to allow microcontrollers and devices to communicate with each other's applications without a host computer.
CAN Controller area network
CCTV Closed-circuit television
CDMA Code-division multiple access
CEP Complex event processing
CFT Crash fault tolerance
Cloud computing The on-demand availability of computer system resources, especially data storage (cloud storage) and computing power, without direct active management by the user.
CO_2 Carbon dioxide
CPMS Car park management system
CPOS Car park operating system
CQELS Continuous query evaluation over linked streams
CQELS-QL CQELS query language
C-SPARQL Continuous-SPARQL
DBMS Database management system
DCPL Direct current power line; a two-wire, active, direct current, analog signal fieldbus technology.
DDoS Distributed denial of service
DNS Domain name server
DSMS Data stream management systems
EA Evolutionary algorithms
ECA rule Event-condition-action rule
ECG Electrocardiogram
Edge computing A distributed computing paradigm that brings computation and data storage closer to the sources of data.

EHR Electronic health records
EHS Electronic height sensor
EMR Electronic medical records
EPC Electronic product code
EPM Enterprise performance management
EP-SPARQL Event-processing SPARQL
ETALIS An expressive language for specifying and combining complex events.
EU European Union
Fieldbus An industrial network system for real-time distributed control.
FL Fuzzy logic
FLPs Facility location problems
Fog computing A decentralized computing structure located between the cloud and devices that produce data.
GDP Gross domestic product
GDPR General data protection regulation
GPRS General packet radio service
GPS Global positioning system
HDFS A distributed file system that handles large data sets running on commodity hardware.
HDG Healthcare data gateway
Homomorphic encryption A form of encryption that permits users to perform computations on its encrypted data without first decrypting it.
HTTP Hypertext transfer protocol
IaaS Infrastructure-as-a-service
IAPNP IoT-based advanced automobile parking navigation platform
IBFT Istanbul-BFT
IBM International Business Machines Corporation
ICB Intelligent city building
ICF Intelligent Community Forum
ICO Initial coin offering
ICT Information and communications technology
ID Identity document
IESE Cities in Motion Index A study published annually by the business school of the University of Navarra (IESE) that aims to evaluate the development of the world's cities.
IMD Smart City Index The index ranks cities based on economic and technological data, as well as by their citizen's perceptions of how "smart" their cities are.
IoT Internet of things
IP Internet protocol
IPFS Interplanetary file system
IRI Internationalized Resource Identifier
IRP Inventory routing problem
IT Information technology
JBoss Drools An open-source business rule management system. It provides an extensive framework for setting up rule-based systems.
JDBC Java Database Connectivity
JSON JavaScript Object Notation
JSON-LD JSON for linking data

Kbit Kilobit
KPIs Key performance indicators
KSI Kolmogorov-Smirnov importance
LANs Local area networks
LCD Liquid crystal display
LED Light-emitting diode
LTE Long-term evolution
LVMH Moët Hennessy Louis Vuitton
MapReduce A programming model and an associated implementation for processing and generating big data sets with a parallel, distributed algorithm on a cluster.
Mbit Megabit
ML Machine learning
Mph Miles per hour
MQ Message queuing
MQTT Message queuing by telemetry transport
MRI Magnetic resonance imaging
NaaS Network-as-a-service
NFC Near-field communication
NGB Next generation broadcasting
NGN Next generation network
NoSQL A type of database that stores and retrieves data without needing to define its structure first.
NOx Nitrogen oxides
OBUs On-Board Units
OECD Organisation for Economic Co-operation and Development
OWL Web ontology language
P2P Peer-to-peer
PaaS Platform-as-a-service
Parallel processing A type of computation in which many calculations or processes are carried out simultaneously.
PB Petabyte
pBFT Practical Byzantine Fault Tolerance
PoB Proof of burn
PoET Proof of elapsed time
PoI Proof of importance
PoS Proof of stake
PoW Proof of work
PoWeight Proof of weight
Proxy A server application that acts as an intermediary between a client requesting a resource and the server providing that resource.
PSTN Public switched telephone network
QoS Quality of service
R2RML RDB to RDF mapping language
RDBMS Relational database management system
RDBS Relational database system
RDD Resilient distributed data-set
RDF Resource Description Framework

RDF-S RDF-Schema
RFID Radio-frequency identification
RIF Rule Interchange Format
RPi Raspberry Pi, a series of small single-board computers developed in the United Kingdom by the Raspberry Pi Foundation in association with Broadcom.
RSP RDF Stream Processing
RSP4J RDF Stream Processing for Java
RSU Road side units
SaaS Software-as-a-service
SASS/SCSS Smart crane selection system
SEQ The intelligent search, analysis, and alerting server built specifically for modern structured log data.
SHA Secure hash algorithm
SPARQL An RDF query language—i.e., a semantic query language for databases—able to retrieve and manipulate data stored in RDF format.
SPEB Smart power extension bar
SSaaS Sensor-as-a-service
STaaS Storage-as-a-service
T&M Transport and citizens' mobility
TD-SCDMA Time division-synchronous code division multiple access
TEE Trusted execution environment
TGT Towngas Telecom
TripleWave A reusable and generic tool that enables the publication of RDF streams on the web.
Turtle Terse RDF triple language.
UNCTAD United Nations Conference on Trade and Development
URI Uniform resource identifier
URL Uniform resource locator
US United States
USB Universal serial bus
VPN Virtual private network
VRP Vehicle routing problem
W3C World Wide Web Consortium
WCDMA Wideband code division multiple access
Wi-Fi Wireless fidelity
WPS Wi-Fi protected setup
WSAN Wireless sensor and actuator network
WWW World wide web
XML Extensible markup language
YAML Yet another markup language
YARN Yet another resource negotiator

Index

Note: Page numbers followed by *f* indicate figures and *t* indicate tables.

T

FSC
www.fsc.org
MIX
Papier | Fördert gute Waldnutzung
FSC® C083411

Kolibri 360 GmbH
Ferdinand-Jühlke-Straße 7
99095 Erfurt, Deutschland
produktsicherheit@kolibri360.de